Katinka Ruth

(R)-3-hydroxycarboxylic acids from bacterial polyhydroxyalkanoates

Katinka Ruth

(R)-3-hydroxycarboxylic acids from bacterial polyhydroxyalkanoates

Production of (R)-3-hydroxycarboxylic acids
and investigation of the physiological role
of polyhydroxyalkanoate degradation

Südwestdeutscher Verlag für Hochschulschriften

Impressum/Imprint (nur für Deutschland/ only for Germany)
Bibliografische Information der Deutschen Nationalbibliothek: Die Deutsche Nationalbibliothek
verzeichnet diese Publikation in der Deutschen Nationalbibliografie; detaillierte bibliografische
Daten sind im Internet über http://dnb.d-nb.de abrufbar.
Alle in diesem Buch genannten Marken und Produktnamen unterliegen warenzeichen-, marken-
oder patentrechtlichem Schutz bzw. sind Warenzeichen oder eingetragene Warenzeichen der
jeweiligen Inhaber. Die Wiedergabe von Marken, Produktnamen, Gebrauchsnamen,
Handelsnamen, Warenbezeichnungen u.s.w. in diesem Werk berechtigt auch ohne besondere
Kennzeichnung nicht zu der Annahme, dass solche Namen im Sinne der Warenzeichen- und
Markenschutzgesetzgebung als frei zu betrachten wären und daher von jedermann benutzt
werden dürften.

Verlag: Südwestdeutscher Verlag für Hochschulschriften Aktiengesellschaft & Co. KG
Dudweiler Landstr. 99, 66123 Saarbrücken, Deutschland
Telefon +49 681 37 20 271-1, Telefax +49 681 37 20 271-0, Email: info@svh-verlag.de
Zugl.: Zürich, ETH, Diss., 2008

Herstellung in Deutschland:
Schaltungsdienst Lange o.H.G., Berlin
Books on Demand GmbH, Norderstedt
Reha GmbH, Saarbrücken
Amazon Distribution GmbH, Leipzig
ISBN: 978-3-8381-0692-2

Imprint (only for USA, GB)
Bibliographic information published by the Deutsche Nationalbibliothek: The Deutsche
Nationalbibliothek lists this publication in the Deutsche Nationalbibliografie; detailed
bibliographic data are available in the Internet at http://dnb.d-nb.de.
Any brand names and product names mentioned in this book are subject to trademark, brand or
patent protection and are trademarks or registered trademarks of their respective holders. The
use of brand names, product names, common names, trade names, product descriptions etc.
even without a particular marking in this works is in no way to be construed to mean that such
names may be regarded as unrestricted in respect of trademark and brand protection legislation
and could thus be used by anyone.

Publisher:
Südwestdeutscher Verlag für Hochschulschriften Aktiengesellschaft & Co. KG
Dudweiler Landstr. 99, 66123 Saarbrücken, Germany
Phone +49 681 37 20 271-1, Fax +49 681 37 20 271-0, Email: info@svh-verlag.de

Copyright © 2009 by the author and Südwestdeutscher Verlag für Hochschulschriften
Aktiengesellschaft & Co. KG and licensors
All rights reserved. Saarbrücken 2009

Printed in the U.S.A.
Printed in the U.K. by (see last page)
ISBN: 978-3-8381-0692-2

Diss. ETH No. 17884

Production of (R)-3-hydroxycarboxylic acids from bacterial polyhydroxyalkanoates (PHA) and investigation of the physiological role of PHA degradation

A dissertation submitted to the
SWISS FEDERAL INSTITUTE OF TECHNOLOGY ZÜRICH
For the degree of
Doctor of Natural Science

Presented by
KATINKA MEIKE RUTH
M.Sc. Chemistry, Technical University of Munich (D)
Born January 31, 1980
in Marburg, Germany

Accepted on recommendation of
Prof. Dr. T. Egli, examiner
Dr. M. Zinn, co-examiner
Prof. Dr. D. Jendrossek, co-examiner
Prof. Dr. S. Panke, co-examiner
Dr. Q. Ren, co-examiner

St. Gallen, 2008

Table of Contents

Summary		6
Zusammenfassung		8
Chapter 1	General introduction	11
Chapter 2	Bacterial poly(hydroxyalkanoates) as a source of chiral hydroxyalkanoic acids	35
Chapter 3	Efficient production of (R)-3-hydroxycarboxylic acids by biotechnological conversion of polyhydroxyalkanoates and their purification	51
Chapter 4	Process engineering for production of chiral hydroxycarboxylic acids from bacterial polyhydroxyalkanoates	65
Chapter 5	Degradation of polyhydroxyalkanoates enhances alkaline stress tolerance in *Pseudomonas putida* GPo1	75
Chapter 6	Identification of two acyl-CoA synthetases from *P. putida* GPo1: One is located at the surface of polyhydroxyalkanoates (PHA) granules	83
Chapter 7	Investigation of an acyl-CoA synthetase knockout mutant of *P. putida* GPo1	97
Chapter 8	General discussion	107
References		
Appendix		130

Summary

Pseudomonas putida GPo1 can accumulate polyhydroxyalkanoates (PHA) as carbon and energy reserve when carbon is in excess and other vital nutrients are limited. PHA is a polyester derived from condensation of activated hydroxycarboxylic acids. Hydrolysis of PHA leads to enantio-pure hydroxycarboxylic acids, which are laborious to prepare by organic syntheses. In this study, *P. putida* GPo1 was exploited to produce various *R*-3-hydroxycarboxylic acids (HA). Being multi-functional chiral syntons, HA are valuable starting materials for synthesis of pharmaceuticals, vitamins, flavors, pheromones, or antibiotics. Since approximately 150 HA have been identified in bacterial PHA, the biotechnological processes developed in this thesis provide a potential new route for the production of various enantiomerically pure chemicals.

An environmentally friendly and highly efficient way to produce different HA has been accomplished. The process developed in this thesis involves PHA accumulation, e.g., under dual-nutrient-limited growth in chemostat culture, and subsequent *in vivo* depolymerization to HA. An alkaline pH is crucial for optimal *in vivo* depolymerization in order to enhance PHA depolymerase activity and excretion of HA into the medium. As an example, we produced eight different HA, including HAs with terminal double bonds. This practical method for the production of HA is described in **chapter 2**.

Producing a mixture of different HA requires a subsequent separation for obtaining the enantio-pure chemicals. *P. putida* GPo1 synthesizes PHA from alkanes or alkanoates via the β-oxidation pathway. Thus, the precursor compounds built into PHA consist of the alkan(oates) with the original chain length and β-oxidation degradation products that are shortened by one or more C_2 units. Monomers obtained from PHA consequently are a mixture of HA whose side chains differ in the length of ethylene subunits. An efficient method for separation and purification of these compounds was developed involving column chromatography and solvent extraction. The process provides pure HA in high yields suitable as fine chemicals for industrial applications. It is environmentally friendly because the used solvents can be recycled. **Chapter 3** presents the approach to isolate several HA, produced by *in vivo* depolymerization, from the culture broth.

To further improve the efficiency of HA production from chemostat cultured cells from *P. putida* GPo1, a continuous *in vivo* bioprocess was designed by exploiting both, its PHA-synthesizing and its PHA-degrading ability. The setup includes a chemostat culture for controlled PHA accumulation from where the cells are transferred into a second reactor where the pH is shifted into the alkaline range. This process design forces HA release to the medium. The number of steps in the procedure was reduced by four, e.g., rendering unnecessary tedious resuspension of cell pellets, which probably will simplify the scale-up of HA production. In combination with high cell density cultivation, this novel bioprocess is an economical and environmentally friendly way to produce chiral HA at a high degree of purity for further applications. This procedure was patented (patent application PCT/CH2007/000156) and is described in **chapter 4**.

Summary

In the next three chapters, the process of HA production by *in vivo* depolymerization of bacterial PHA was investigated with respect to its metabolic background. Regarding underlying mechanisms, the physiological advantage of HA excretion for the bacterial cell itself was investigated. Under alkaline conditions, *in vivo* depolymerization of PHA provides a way to neutralize pH in direct cellular surrounding. Thus, HA release enhances stress tolerance and cellular survival. The intracellular pH was maintained to a certain extent in PHA-degrading cells. A hypothesis for pH homeostasis and alkaline stress tolerance is proposed in **chapter 5**, involving a mechanism controlled by either enzymatic activities or genetic regulations of the general stress response.

In order to elucidate PHA degradation pathways in *P. putida* GPo1, PHA granule-associated proteins were examined. An acyl-coenzyme A synthetase (ACS1) was discovered and its *in vivo* localization at the surface of PHA granules was confirmed by fluorescence microscopy studies with ACS-fusions with the green fluorescent protein. ACS1 is assumed to activate HA with CoA for further utilization by the cells as a carbon and energy source. A second ACS (ACS2) was cytosolic or partially associated with cellular membranes and did not seem to be directly involved in PHA metabolism. Properties of the two ACS including their *in vivo* localizations are presented in **chapter 6**.

An ACS1 knockout mutant of *P. putida* GPo1 was constructed to block the metabolic re-utilization of HA by preventing their activation with coenzyme A. When grown on fatty acids, the mutant had a prolonged lag phase, which might indicate that ACS2 has to be induced first and then can assume the role of ACS1. Preliminary data showed that the mutant accumulated less PHA and its PHA degradation was not impaired. Studies with this ACS deletion mutant allowed new insights into the role of ACS in PHA metabolism; these are described in **chapter 7**.

Hence, this thesis demonstrates an efficient method to produce HA from bacterial PHA and provides new details about PHA degradation in *P. putida* GPo1.

Zusammenfassung

Der Mikroorganismus *Pseudomonas putida* GPo1 kann Polyhydroxyalkanoate (PHA) einlagern, wenn Kohlenstoffsubstrate im Überfluss vorhanden sind und gleichzeitig ein anderer lebenswichtiger Nährstoff das Wachstum limitiert. PHA ist ein Polyester, der durch die Kondensation von aktivierten Hydroxycarbonsäuren entsteht. Die Hydrolyse von PHA führt zu enantiomerenreinen Hydroxycarbonsäuren, die mittels chemischer Synthese nur schwer herzustellen sind. In dieser Studie wurde *P. putida* GPo1 zur Herstellung verschiedener. R-3-Hydroxycarbonsäuren (HA) benutzt. HA sind multifunktionelle chirale Synthons und damit wertvolle Ausgangsstoffe für die Synthese von Arzneimitteln, Vitaminen, Aromen, Pheromonen oder Antibiotika. In bakteriellem PHA wurden bisher rund 150 HA identifiziert. Somit ermöglicht der in dieser Arbeit entwickelte biotechnologische Herstellungsprozess vorrausichtlich einen neuen Weg zur Darstellung vieler enantiomerenreiner Feinchemikalien.

Als umweltfreundlicher und effizienter Ansatz zur Herstellung von HA erwies sich die Anreicherung von PHA in Bakterien, zum Beispiel durch deren Kultivierung in doppelt Nährstoff limitierten Chemostaten, und eine darauffolgende PHA Depolymerisation zu HA in lebenden Bakterienzellen. Für eine optimale *in vivo* Depolymerisation ist ein alkalischer pH-Wert nötig, der die PHA-Depolymerase Aktivität erhöht und das Ausschleusen der HA ins Medium verstärkt. In dieser Studie wurden exemplarisch 8 HA (einschliesslich Verbindungen mit endständigen Doppelbindungen) hergestellt. Die dazu entwickelte praktische Methode ist in **Kapitel 2** beschrieben.

Um Reinstoffe zu erhalten, muss das Gemisch aus verschiedenen HA einem Trennungsvorgang unterworfen werden. *P. putida* GPo1 wandelt Alkane und Fettsäuren über die β-Oxidation in PHA um. Darum können die PHA Vorstufen um ein oder mehrere C_2-Körper verkürzt werden, sodass PHA aus einem Copolymer besteht, dessen Seitenketten sich um die Länge von ein oder mehreren Ethylen-Untereinheiten unterscheiden. Die aus PHA gewonnen Monomere spiegeln diese Verteilung in ihrer Zusammensetzung wider. Zu ihrer Trennung und Reinigung wurde eine effiziente Methode entwickelt, die auf Anwendung von Säulenchromatographie und Lösemittel Extraktion beruht. Mit ihr können HA als Reinstoffe für industrielle Anwendungen gewonnen werden. Alle Lösemittel können wiederverwendet werden, was die Methode umweltfreundlich macht. Die Methode zur Isolation verschiedener HA aus einer Bakterienkultur wird in **Kapitel 3** vorgestellt.

Um die Effizienz der HA Herstellung aus Chemostatkulturen von *P. putida* GPo1 weiter zu erhöhen, wurde ein kontinuierlicher Bioprozess entwickelt, der die Fähigkeit von *P. putida* GPo1 ausnutzt PHA auf- und abzubauen. Der Aufbau umfasst einen Chemostaten zur kontrollierten PHA Anreicherung und einen daran angeschlossenen zweiten Reaktor in dem der pH-Wert im alkalischen gehalten wird, was eine Ausscheidung der HA ins Medium erzwingt. Durch diesen Bioprozess wird die Zahl der Arbeitsschritte reduziert, was ein Maßstabsvergrößerung der HA Herstellung erleichtern könnte. In Kombination mit Hochzelldichten ist dies ein wirtschaftlicher und umweltfreundlicher Weg chirale HA mit hohem Reinheitsgrad zu gewinnen. Der Prozess wurde patentiert (PCT/CH2007/000156) und wird in **Kapitel 4** erläutert.

Zusammenfassung

In den folgenden drei Kapiteln wurde der metabolische Hintergrund untersucht, der der HA Herstellung durch Depolymerisation von bakteriellem PHA zugrunde liegt. Der physiologische Vorteil des HA-Ausschleusens für die Bakterienzelle wurde ermittelt. Im alkalischen bietet dieser Mechanismus einen Weg die direkte Umgebung zu neutralisieren. Die HA-Ausscheidung dient also der Stresstoleranz und dem Überleben. In PHA abbauenden Zellen kann der intrazelluläre pH bis zu einem gewissen Grade aufrecht erhalten werden. Ein möglicher Mechanismus der alkalischen Stresstoleranz und der pH-Homeostasis in *P. putida* GPo1, bei dem pH-abhängige Enzymaktivitäten oder die genetische Regulationen der allgemeinen Stressantwort eine Rolle spielen, wird in **Kapitel 5** erläutert.

Um die Abbauwege vom PHA besser zu verstehen, wurden die Proteine an den intrazellulären PHA Einschlüssen untersucht. Eine Acyl-Coenzym A synthetase (ACS1) wurde entdeckt und ihr Auftreten an der Oberfläche der PHA Einschlüsse durch Fluoreszenzmikroskopiestudien mit Bakterien bestätigt, die Fusionen aus ACS und Grün fluoreszierenden Protein enthielten. Vermutlich aktiviert ACS1 die HA mit CoA, um sie der Zelle als Kohlenstoff- und Energiequelle zugänglich zu machen. Eine zweite ACS (ACS2) befindet sich im Zellplasma und zum Teil an der Zellmembran. Dieses Enzym schien nicht direkt am PHA Metabolismus beteiligt zu sein. Eigenschaften der zwei ACS, insbesondere ihr Aufenthaltsort innerhalb der Zelle, werden in **Kapitel 6** vorgestellt.

Um die Aktivierung von HA mit CoA zu blockieren und so deren Verwertung im Stoffwechsel zu unterbinden, wurde eine ACS1 Knockout Mutante von *P. putida* GPo1 entwickelt. Das Wachstum der Mutante zeigte eine verlängerte Lag-Phase, wenn Fettsäuren als Kohlenstoffsubstrat zugegeben wurden. Das deutet vermutlich darauf hin, das ACS2 zuerst induziert wird und dann die Rolle von ACS1 übernehmen kann. Erste Experimente zeigten, dass die Mutante weniger PHA einlagert und dass der PHA Abbau nicht beeinträchtigt ist. Studien mit dieser Mutante erlauben neue Einblicke in die Rolle der ACS beim PHA Stoffwechsel; diese werden im **Kapitel 7** beschrieben.

Zusammenfassend zeigt diese Arbeit eine effiziente Methode zur Herstellung von HA aus bakteriellem PHA auf und liefert neue Erkenntnisse über den PHA Abbau in *P. putida* GPo1.

Chapter 1

General Introduction

The prevention of environmental pollution is one of the most urgent challenges of modern society. Wherever men settled, especially in fast developing countries, ecological balances are disturbed and nature is negatively affected. The amount of plastic waste floating nowadays in the pacific equals the area of central Europe (~26 million square kilometers, 3.5 million tons).[1,2] Many more toxic and often less visible man-made substances are threatening ecological balances and accelerate the extinction of species. Polymer plasticizers, for example, can inhibit propagation and social behavior of tuna and other fish in extremely low concentrations (0.5µg L^{-1}).[3] Scarcity of raw materials and resources together with a growing world population further call for the introduction of new sustainable materials and ecological production processes. Materials from crude oil and petroleum-based plastics have polluting potential during both fabrication and decomposition. Hence, focusing on biotechnological production of sustainable compounds such as biodegradable plastics is certainly a step in the right direction.

Other societal challenges are the high standard of living and obsolescence in developed countries coming along with a demand for innovative medication, e.g., new antibiotics. The biotechnological use of micro-organisms and their enzymes as catalysts enables the synthesis of a wide variety of interesting compounds, which are often not easily accessible with pure synthetic approaches.

Biocatalysis and chiral synthons in industry

Biotransformations are enzyme-catalyzed often environmentally friendly, regio- and stereoselective processes.[4] For many biological and medical applications only one enantiomer is active. When operating with racemates, which nowadays is still common use, production efficiency can maximally reach 50% of active compound. Higher doses are need to administer if the racemate is applied and only one enantiomer is active.[5,6] Economic incentives of enhanced biological activity and improved cost efficiency, together with new regulatory pressures have brought about a dramatic increase in the number and scope of industrial processes utilizing biocatalysis.[7] Therefore, the production of pure enantiomers is a topic of ever-increasing importance to chemical industry.

The replacement of conventional chemical processes by sustainable biotechnological processes is one of the main current tendencies in biocatalysis, white biotechnology and chemical industry.[8] Among others, the synthesis of β-lactam antibiotics has been rendered more efficient by the introduction of one or more enzyme-catalyzed steps. For example, Cefalexin, a first generation cephalosporin produced at a 3000t/a scale, had a synthesis involving ten, largely classical steps which could be replaced by one biotransformation.[9]

In many processes, the combination of methods from organic synthesis and biocatalysis has proven to be very fruitful. Chiral building blocks from natural resources have been used as starting materials to build up complex structures using classical synthetic methods.[10,11]

Biological polymers

Nowadays, polymers have applications in all areas of life and nearly every industry. Polymers are not only produced synthetically by organic chemistry but also plants, animal cells and many prokaryotic microorganisms including eubacteria and archea are able to synthesize them. Sustainable alternatives to petrochemical-derived products have been developed from many biodegradable polymers such as polyhydroxyalkanoates (PHA), polyglycolic acid, polycaprolactone or polylactic acid (PLA). So far, the most commercialized one is PLA, a biodegradable, thermoplastic, aliphatic polyester derived from renewable resources. It is produced from corn starch of sugarcanes via bacterial fermentation and subsequent chemical polymerization. PLA has applications as food packaging and disposable plastics of everyday life as well as biomedical material and in tissue engineering.

Even though a wide variety of biopolymers has been discovered, the most versatile and promising polyesters are PHA.[12] Numerous microorganisms synthesize them and biotechnological cultivation can be used for industrial production. Many substances can serve as starting material such as renewable or fossil resources or even toxic waste products.

Polyhydroxyalkanoates (PHA)

PHA are water insoluble polyesters, which are accumulated by various microorganisms as an intracellular carbon and energy reservoir.[13, 14] They contain a great variety of repeating units, including even uncommon functional groups when the PHA-accumulating bacteria were grown on substrates derived from crude oil or organic syntheses. Microbial polyesters can be easily produced in fermentation processes. The linear polyesters are biodegradable and biocompatible.[14] When exposed to soil bacteria, e.g., in a compost, PHA is degraded to water and CO_2, or to methane.[15] PHA synthesis has been detected in many archea and eubacteria under aerobic as well as anaerobic conditions and has been transferred into eukaryotic microorganisms,[16] recombinant PHA-negative bacteria,[17] animal cells,[18] and plants.[19] Transformation of raw materials or waste products into PHA or PHA derivatives with added value can be considered an important input in terms of eco-effective applicability and environmental biotechnology.[20]

Structures and material properties

Lemoigne was the first to discover polymers in microorganisms, when he isolated poly((R)-3-hydroxybutyrate (PHB) from *Bacillus megaterium*. PHB is an isotactic polyester having all methyl groups in absolute *R*-configuration (Figure 1.1). Molecular weight of PHB *in vivo* is generally between 10^5 and 10^6 depending on the organism.[21, 22]

Only half a century later, Wallen et al.[23] characterized polyesters from microorganisms in sewage sludge containing monomer units with more than four carbon atoms, i.e., longer than (*R*)-3-butyric acid

(HB) from PHB. To date, more than 150 repeating units have been found to be incorporated into PHA and it became clear that a large number of microorganisms are capable of accumulating PHA.[24] Various functional groups have been discovered at the side chains of PHA (Fig. 1.1, see also appendix where a list of PHA monomers and their current production methods is presented).

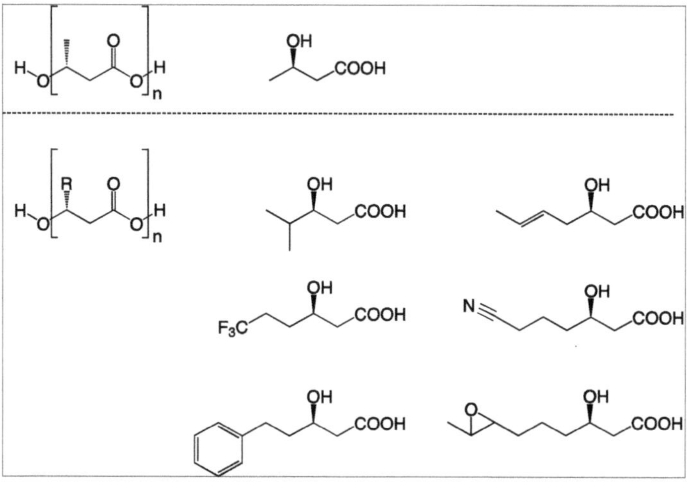

Figure 1.1 PHB and its monomer (R)-3-hydroxybutyric acid (upper panel). PHA and representative monomers, i.e., hydroxycarboxylic acids (HA) (lower panel). HA have a hydroxyl group in 3-, 4-, or 5-position; they can have branched or unsaturated side chains and can contain cyano, halogene, aryl, phenoxy or further hydroxyl or carboxy groups.[13, 24]

In general, structures of PHA are mainly determined by the producing microorganisms and the carbon source supplied as a growth substrate. Structurally, PHA can be classified into three groups based on the number of carbon atoms in the monomer units:[13, 21] Short-chain-length, (scl) PHA, which consist of 3-5 carbon atoms, medium-chain-length, (mcl) PHA, which consist of 6-14 carbon atoms, and long-chain-length, (lcl) PHA, which consist of more than 14 carbon atoms. Concerning polymer characteristics, molecular weight of PHA ranges from $2 \cdot 10^5$-$3 \cdot 10^6$ Da and their melting transition temperatures decrease with increasing length of the side chains and with decreasing homogeneity of the repeating units.[12, 25] Incorporation of functional groups into PHA complicates the prediction of material properties, e.g., introducing double bonds at the side chains also decrease melting and glass transition temperatures, leading to softer and stickier materials. Characteristics like range of glass transition or melting points and crystallinity are determined by constitution (e.g., branching) and functionalization (e.g., substituted by epoxygroups or halogens) of monomeric side chains. Furthermore, material properties strongly depend on culture conditions and are also influenced by down-stream processes and the age of the sample.

To illustrate ubiquity and diversity of PHA, an impressive example is poly-β-malate with a carboxylic function as pendant group, i.e., L-malic acid as monomer unit.[26] The ester bonds are formed in a regio- and stereospecific manner. Poly-β-malate has been found as extracellular, water-soluble biopolyester from *Aureobasidium sp.* as well as within cells of certain fungi (e.g., *Penicillium cyclopium*) and

General introduction

slime molds (e.g., *Physarum polycephalum*). Material properties of poly-β-malate are a molecular weight of 4·10⁴-6·10⁴ Da, a polydispersity of 1.5-3.0 and no glass transition or melting point before thermal decomposition at ~185°C. Its bio-adsorbability and biodegradability makes this material suitable for pharmaceutical applications, e.g., as matrix for drug carriers or artificial bones. Polythioesters are another subgroup with very promising properties, which are currently investigated be Steinbüchel and co-workers.[27, 28]

In biotechnological processes, a huge variety of PHA have been produced during last decades. Outstanding examples are the production of terpolymer from HB, 3-hydroxyvaleric acid (HV), and 5-hydroxyvalerate from *R. eutropha* grown with mixtures of 5-chloropentanoic acid and pentanoic acid,[29] or PHA containing pivalic acid from *Rhodococcus ruber* NCIMB 40126,[30] or tailor-made PHA containing a defined amount of double bonds at the side chains which provide options for further functionalization.[31] Despite the high number of different PHA, only PHB and copolymers of HB and HV have been commercialized so far (e.g., by Biopol®).[32]

Molecular genetics of PHA metabolism

Organization of operon

Many genes related to PHA biosynthesis from various bacteria have been cloned, including not only PHA polymerases, but also other PHA granule-associated proteins, proteins with regulatory functions, and enzymes catalyzing the formation of the PHA precursor's, i.e., hydroxyacyl-CoA thioesters.

The structural genes of PHA metabolism often form a *pha* cluster. In some bacteria even structural genes for processing precursors, such as β-ketothiolase and acetoacyl-CoA reductase are closely located.[33] In *Cupriavidus necator* H16 (formerly *Ralstonia eutropha*), the PHB synthetic genes (*phbCAB*) are organized in a single operon with a transcription start site (promoter) at 307 base pairs (bp) upstream of *phbC* (Fig. 1.2A).[34] Seven additional PHB depolymerase genes, which are not closely located to the PHB synthetic operon, have been identified in *C. necator*.[35]

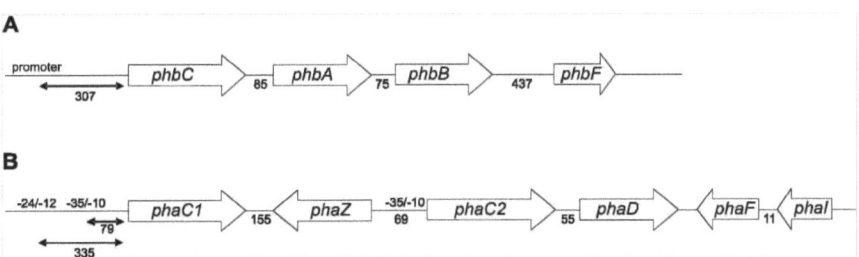

Figure 1.2 A. The *phb* operon of *C. necator* H16 contains a PHB polymerase (*phbC*, 1769 bp), a β-ketothiolase (*phbA*, 1181 bp), an acetoacetyl-CoA reductase (*phbB*, 740 bp) and a regulator (*phbF*, 551 bp).[36] B. The *pha* operon of *P. putida* GPo1 (formerly *P. oleovorans* GPo1) comprises two PHA polymerase genes (*phaC1*, 1677bp and *phaC2*, 1730 bp), one depolymerase (*phaZ*, 849 bp), a regulatory gene (*phaD*, 615 bp) and two phasins with structural function (*phaF*, 767 bp and *phaI*, 419 bp). Mode of operation of promoters (-24/-12 bp, -35/-10 bp) is still under investigation. Numbers indicate amount of bp.[17, 37]

In pseudomonads, two different PHA polymerase genes (*phaC1* and *phaC2*) are found closely related in the genome and with the same orientation (Fig. 1.2B).[17] They are separated by an intracellular PHA depolymerase gene (*phaZ*), located with opposite direction between the two polymerase genes.[38] Genes with structural and regulatory functions in PHA metabolism are often found closely located to PHA synthesis genes.[17, 37]

Regulation of PHA accumulation and degradation

Intracellular storage of PHA acts as a carbon buffer and facilitates survival of bacteria in absence of exogenous carbon sources.[15] For intracellular PHA, a simultaneous process of synthesis and degradation of PHB was confirmed by ^{14}C-glucose pulse experiments.[39-41] The process might depend on intracellular concentrations of PHB-related metabolites or co-factors such as NADH, NAD, or CoA.[42] Phasins may also contribute to regulation of PHA synthesis and decomposition.[43] PHB accumulation is only observed under nutrient-limited growth because under optimal growth conditions PHA polymerase is inhibited by free CoA[44] and because formation of the PHB precursor acetoacetyl-CoA is reversible with the equilibrium on the side of acetyl-CoA.[45]

In *Pseudomonas putida* KT2442, PhaZ is also expressed during with PHA synthesis, however, when these cells were exposed to non-nitrogen limiting conditions, *phaZ* expression was promoted and PhaZ concentration have been fount to be even four times higher.[46] Limitation of vital nutrients might direct a regulatory system for transcription of *phaZ*, as suggested for *P. putida* CA-3.[47] Transcriptional regulation is under investigation: For example, a regulatory gene *phaR* of *P. lemoignei* has been identified with a putative helix-turn-helix motif which might indicate a DNA-binding protein which might be involved in the regulation of PhaZ expression.[48]

In *P. aeruginosa*, two promoters are located upstream of *phaC1* and one promoter upstream of *phaC2*.[49] Co-transcription of *phaC1*, *phaZ* and *phaC2*, or *phaC1* and *phaZ* were reported, but not confirmed.[49, 50] The exact mechanism of how transcription is activated and how PHA accumulation is triggered when growth becomes nutrient-limited has not yet been described. Even though PHA can be produced in larger amounts not many details are known about the underlying mechanisms including the regulation of mcl-PHA accumulation and degradation.[51]

Biosynthesis of PHA

Although there are some *in vitro* approaches to synthesize PHA,[14, 52] the *in vivo* production of PHA is applied more often. For the formation of PHA, different natural organisms such as *C. necator*, *Methylobacterium extorquens* or various strains of *P. putida* as well as recombinant organisms have been used.[36] Growth conditions are generally very diverse and depend, among other things, on production organism, substrates, and the PHA of interest. For PHA accumulation, one nutrient (commonly nitrogen) must limit growth of the biomass while all other nutrients, in particular carbon, have to be in excess.[14]

Different methods of cultivation are feasible:[14, 53, 54] In batch cultures, cells go through a sequence of growth stages[55] due to changing nutrient concentrations caused by cellular metabolism. As soon as one nutrient becomes limiting and carbon is still available, cells start to accumulate PHA. A disadvantage of batch cultivation is the possibility of side-product toxicity (e.g., acetate). Insufficient oxygen supply may impair cell growth. Such growth conditions rule out the possibility to obtain homogeneous polymers and to reach maximal PHA accumulation. Nevertheless, batch cultures are technically simple, can reach high cell densitites, and well suited for growth studies and screening for potential PHA-accumulating organisms.[14]

In fed-batch cultivations, nutrients are added as soon as one of the required medium components gets depleted in the preceding batch culture.[55] Fed-batch cultivation has been successfully used on a large scale in industry since high cell densities can be achieved.[56, 57] Disadvantages of fed-batch cultures are the decreasing growth rate of cells due to linearly increasing cell concentration when the feed rate is kept constant. This may cause a shift in copolymer composition and losses in PHA productivity.[58, 59] Exponential feeding rates are convenient for *P. putida* KT2442, but not feasible for *P. putida* GPo1 which requires nitrogen limitation for PHA accumulation.[60] Larger bioreactors are needed when compared with continuous culture.

Continuous culture is the most controlled cultivation method.[14, 53, 54] The culture broth is continuously harvested and exchanged with sterile growth medium[55] at a fixed dilution rate (D), which is defined by the ratio of feed rate to culture volume.[53] The continuous culture is a chemostat culture when steady-state is reached, i.e., when deviations of all variables as a function of time are zero.[53] When steady-state growth conditions are established, the specific growth rate µ equals D and all concentrations remain constant. PHA accumulation can then be controlled by D and the ratio of carbon source to limiting nutrient.[53, 61] Simultaneous limitation of growth by two heterologous nutrients (e.g., N and C) under chemostat culture conditions, thus a dual-nutrient-limited growth regime[62, 63] enables optimal PHA production.[64] Growth limiting nutrients can be O, N, P, Mg, K, or S. The regime of dual-nutrient-limited growth depends on the ratio of the two limiting nutrients and the specific growth rate of the cultivation.[65] The controlled carbon feed can be optimally adjusted, which might be crucial if too high substrate concentration are toxic (e.g., some alkanoic acids and short chain alkanes). The well-defined growth conditions in a chemostat facilitate investigations on cell physiology and the production of homogenous PHA polymers.[14]

Besides using cultivation of microorganisms, production of PHB, PHBV and mcl-PHA has also been accomplished in transgenic plants such as mutants of *Arabidopsis thaliana* and *Brassica napus*. Larger scale and more competitive production costs are major incentives for PHA synthesis in plants, however, further metabolic engineering is needed to control the polymer composition and increase production yields.[66]

A very promising producer strain is *P. putida* GPo1 which supports mcl-PHA production by alkanes and carboxylic acids having more than 5 carbon atoms, whereas smaller substrates lead to increased cell growth rather than PHA production.[67] In general, PHA production by biosynthesis is well established; however, it is essential to reduce production costs, especially those of down-stream processing, in order to make PHA available on the market at a reasonable price.

Applications of PHA

Biopolyesters are very interesting because of their biocompatibility, biodegradability, and permeability. Mcl-PHA are of increasing interest to industry because of their broad range of material characteristics. Potential applications of mcl-PHA are biodegradable plastics,[68] materials in medicine and pharmaceutical industries,[69] and sources of chiral monomers.[70]

PHA could replace petrochemical polymers currently omnipresent in daily life (e.g., as packaging or coating materials), since they exhibit similar material properties as conventional plastics.[32] PHA have an enormous potential for medical applications being biodegradable (e.g., polyhydroxyoctanoate films degrade in saline within ~2 years[71]) and biocompatible, i.e., they do not cause inflammations when implanted in living animals. PHA have been investigated as drug carriers[72, 73] or as scaffold material in tissue engineering.[74] Moreover, the use of PHA as skin substitutes, cardiovascular fabrics, bone graft substitutes or internal fixation devices (e.g., screws) has also been considered.[14]

In contrast, PHB is a polymer which is too stiff and brittle for most commodities.[75] This is due to its high degree of crystallinity, its highly spherulite morphology, and its melting point (180°C) close to its thermal degradation temperature (200°C). Compared to scl-PHA, mcl-PHA are less crystalline, thus they are more flexible and less brittle thermoplasts and they have lower glass transitions and melting points. By choosing appropriate side chains, mcl-PHA can be tuned to exhibit desired material properties (tailor-made synthesis), which makes this material much more attractive for applications.

However, even these materials sometimes need chemical modification to improve their mechanical and viscoelastic properties for some medical and industrial applications. Therefore, physical properties of the polymer can be modified by chemical functionalization such as copolymerization, branching, or cross-linking.[76] For example, cross-linking of unsaturated side chains in mcl-PHA increases hardness and tensile strength of the material.

Having different types of 3-, 4-, 5- or 6-hydroxyalkanoic acids with an absolute (*R*)-configuration as monomer units,[36, 77] PHA displays a unique potential as a source of R-3-hydroxyalkanoic acids and derivatives thereof.

PHA monomers

Structures of PHA monomers

Up to date, about 150 PHA monomers have been identified.[13, 24, 78] Not only 3-hydroxy, but also 4-, 5-, or 6-hydroxycarboxylic acids have been found to be incorporated into bacterial polymers and new compounds are still being discovered. All PHA monomers show absolute (R)-configuration, they are enantiomerically pure.[79] Their side chains can contain double bonds as well as more exceptional functional groups such as halogens or phenoxygroups.[13] A survey of most interesting (R)-3-hydroxycarboxylic acids (HA) is given in appendix 1. HA are valuable chiral synthons and can be widely used as chiral building blocks for synthesis of antibiotics, pharmaceuticals, vitamins, flavors and pheromones.[80, 81]

Current production methods of PHA monomers

The defined introduction of the chiral centre into the carbon chain is the challenging step when producing enantiomerically pure HA. Different approaches have been reported. Classical organic synthesis may include stereoselective oxidation through Sharpless' asymmetric epoxidation and consecutive hydroxylation or through Brown's asymmetric allyboration.[82] Enantiopure 3-hydroxyesters have been prepared chemically via enantioselective reduction of 3-keto esters as prochiral precursors.[83, 84] General drawbacks of these reactions often are the requirement of chiral, often expensive metal-complex catalysts and pure substrates. Vigorous reaction conditions such as high pressure, flammable reaction media, or cryogenic conditions are needed[82, 85] and the range of possible products is limited. The necessity to synthesize precursor molecules may complicate the synthetic procedure and may reduce the product yield.[86, 87] The main disadvantage is a lower enantiomeric excesses compared to biochemical processes.

To circumvent these problems, other approaches include microorganisms as biocatalysts to introduce the chiral center.[80, 88] Commonly used strains are recombinant *Escherichia coli*[89] or *Saccharomyces cerevisae*.[90, 91] However, product inhibition of enzyme activity and moderate yields of product isolation can be limitations of these procedures.[70, 81] Additionally, it requires expensive starting compounds and complicated experimental setups, which lead to high production costs, and in some cases even enantiomeric excess can be suboptimal.

Alternatively, bacterial PHA can be used as the source of enantio pure HA. Methods involving acid catalyzed hydrolysis of the extracted PHA have been reported.[92, 93] However, large amounts of organic solvents were used and the production efficiency was rather low due to multistep processes.[92]

An attractive approach to obtain HA from PHA is the *in vivo* depolymerization. The process utilizes intracellularly located PHA depolymerases for hydrolysis of PHA.[15, 94-96] It has been efficiently accomplished with naturally PHB-accumulating bacteria to produce (R)-3-hydroxybutyrate (HB) with a yield of 96% (w/w).[70] Appropriate environmental conditions are crucial for this process. Lee et al. reported that with *Alcaligenes latus*, lowering the pH to 3-4 induced the highest activity of intracellular PHB depolymerase and blocked the reutilization of HB by the cells.[70] In order to gain access to more interesting molecules, *in vivo* depolymerization of mcl-PHA in *Pseudomonas* has been studied as well, however, the production efficiency of mcl-HA was very low (maximum 9.7% after 4 days).[70] Metabolism of mcl-PHA is not as well studied as the one of scl-PHA and involves different reaction pathways; that is why further investigations are necessary to improve this process.

Chiral HA are attractive compounds with high potential. Nevertheless, only few of these compounds are commercially available, such as (R)-3-hydroxybutyric acid, or molecules having no stereocentre or less than four carbon atoms. So far, only two enantiomerically pure HA with more than 5 carbon atoms are placed on the market: (R)-3-hydroxynonanoic acid (supplied by Exclusive Chemistry Ltd, Russia) and (R)-3-hydroxytetradecanoic acid (supplied by Wako Pure Chemical Industries Ltd, Japan).

Potential applications of PHA monomers

HA can be used to produce novel polymers with new properties and to synthesize β- and γ-amino acids or peptides.[97] Since β-peptides are stable to peptidases, they can be used as scaffolds for peptide mimics.[98] They also can serve as mono-functional spacers for peptide-related drugs. For example, several non-peptidic spacers have been designed which replace antigenic peptides in a MHC-peptide complex (MHC major histocompatibility complex, immunology). Thus, T-cell response to the parent antigenic peptide can be manipulated.[99] 3HB oligomers provide energy and show good penetration and rapid diffusion in peripheral animal tissue, hence they could be an energy substrate for injured patients.[100]

Chiral synthons for organic synthesis

Since HA are not readily available on the market and their classical synthesis is rather tedious, only few methods using such reactants as starting material have been reported. Many syntheses use HA as reactants and could be carried out in a more efficient and strikingly shorter way when applying HA produced by *P. putida* GPo1. The following examples are given to illustrate the versatility of syntheses using HA as reactants.

The α,β-unsaturated δ-lactone (R)-massoialactone is a constituent of native medicine to cure various infections for many centuries.[101] It shows antimicrobial and antifungal activity and has been isolated from *Cryptocarya massoia* as well as from jasmine flowers.[102] Like many aliphatic δ-lactones it occurs in several food flavours and essential oils which play an important role as flavouring materials, due to their specific odour impression and low threshold concentration. Touati et al.[103] succeeded in the synthesis of massoialactone using a ruthenium-arylphosphine catalyst [RuBr$_2$((R)-SYNPHOS)] to produce the intermediate (R)-3-hydroxyoctanoate through asymmetric hydrogenation. Chiral synthesis of another δ-lactone, 3,5-dihydroxydecanoic acid, using (R)-3-hydroxyoctanoate has also been reported.[104]

Both synthesis use rather expensive complex metal catalysis to introduce the chiral centre. Hence, easy accessibility of HA would simplify these two and many other syntheses of saturated and unsaturated aliphatic δ-lactones, which are often interesting bioactive natural products.

Gloeosporone is a spore germination autoinhibitor, containing a 14-membered macrolide. (R)-3-hydroxyoctanoate methyl ester can be used as starting substance.[105] In this multistage synthesis, the method of choice for producing this compound is an asymmetric reduction of β–keto esters to cope with stereochemical problems. A Ru(II)-BINAP catalyst has been developed for this purpose.[83] Challenging metal-catalyzed reactions can be shirked when starting with (R)-3-hydroxyoctanoate from bacterial source.

Certain divinyl ether fatty acids inhibit mycelia growth and spore germination in certain fungi.[106] HA can serve as chiral precursors for studies with those compounds and, e.g., (R)-3-hydroxyheptanoate was reported to be obtained by several crystallization steps from carcinogenic carbon tetrachloride for this purpose.[107]

Linear condensed triquinane sesquiterpenes are constituent of essential oils in plants and of great economical interest. Starting materials such as bicyclo[3.2.0]hepte-3-en-6-ones[108] or tetrahydro-2H-cyclopenta[b]furan-2-ones[109] can be synthesized with (R)-3-hydroxyhept-6-enoate as reactant. (R)-3-hydroxyhept-6-enoate is also a precursor for eremophilan carbolactones[110] which are valuable compounds for drug design and also known in traditional Chinese medicine. They showed positive effects on blood circulation and rheumatism and some were found to exhibit antimicrobial activities.[111]

In fact, there are a lot of bioactive and pharmaceutical interesting molecules known containing HA as substructures. Hence, the accessibility of these chiral acids as building blocks might open new synthetic routes towards many important compounds. Table 1 gives an idea about the versatility of applications of PHA monomers.

Table 1. Potential applications for selected PHA monomers.

PHA monomer	Potential synthon for	Reference
(R)-3-hydroxyundec-10-enoate	Inhibitor of cholesterol synthesis, effects 3-hydroxy-3-methyl-glutaryl (HMG) CoA synthetase	Dirat et al.[112]
	Precursor of L-659,699 (inhibitor of cholesterol biosynthesis)	Chian et al.[113]
(R)-3-hydroxy-undecanoate	Depsipeptides (antibiotic/antifungal)	Wohlrab et al.[114] Nihei et al.[115]
	(-)-tetrahydrolipstatin (anti-obesity drug)	Yin et al.[116] Pons et al.[117]
	Lipid A mimic (immunobiological)	Martin et al.[118]
	Stevastelins B and B3	Sarabia et al.[119]
	Sulfobacin A	Gutpta et al.[120] Irako et al.[121] Labeuw et al.[122]
	Globomycin (antibiotic, signal peptidase II inhibitor)	Kiho et al.[123-125]

	Pseudomycin	Rodriguez et al.[126]
	Topostins B567 and D654	Shioiri et al.[127]
(R)-3-hydroxy-nonanoate	Globomycin analogues (antibiotic)	Kiho et al.[125]
	(R)-2-benzylcyclohexanone (precursor of natural products)	Katoh et al.[128]
(R)-3-hydroxy-octanoate	Simvastatin (antihypercholesterolemic, inhibitor of HMG-CoA reductase)	Morgan et al.[129] Lee et al.[130]
	Viscosin	Hiramoto et al.[131]
(R)-3-hydroxyhept-6-enoate	Sphingofungin D (antifungal)	Mori et al.[132] Vanmiddlesworth et al.[133]
	Rosuvastatin calcium, HMG CoA reductase inhibitor	Zlicar et al.[134]
	α,β-disubstituted β-lactones	Wu et al.[135]
	Potent HMG CoA reductase inhibitor FR901512	Inoue et al.[136]
	Sphingofungin F (antifungal)	Kobayashi et al.[137]
	Precursor of β-lactams for synthesis of carbacephems (class of antibiotics)	Crocker et al.[138]
	Ebelactone A and B (β-lactone enzyme inhibitor)	Paterson et al.[139]
	Bicycloheptenones	Marotta et al.[140]
	Cyclosporine A derivatives (immunosuppressive)	Aebi et al.[141] Colucci et al.[142] Rich et al.[143] Schreiber et al.[144] Schmidt et al.[145] Lynch et al.[146]
	(S)-citronellol	Hirama et al.[147]
(R)-3-hydroxy-heptanoate	Anachelin (siderophore of *Anabaena cylindrica*)	Ito et al.[148]
	Pravastatin (atherosclerosis/hypercholesteremia agent)	Keri et al.[149]
(R)-3-hydroxy-hexanoate	Analogues of laulimalide (paclitaxel like antimicrotubule agent)	Faveau et al.[150]

In Table 1, only HA directly investigated in this thesis have been considered. Taking into account all HA that could be produced biotechnologically, the number of potential synthesis pathways towards pharmaceutically active compounds increases enormously.

Homopolymers and tailor-made block-copolymers

A special application of chiral HA is the possibility to synthesize tailor-made polymers when using them as monomers in a condensation reaction. Since (R)-3-hydroxycarboxylic acids from bacterial source have 100% (R)-configuration, they would result in polyesters composed of isotactic macromolecules. The process of *in vivo* depolymerization of PHA to obtain HA and their consecutive polymerization might seem circuitous. However, it could open a new way to a totally isotactic class of homopolyesters with special characteristics and unique properties. It is also feasible to synthesize block or graft copolymers using several types of different 3-hydroxycarboxylic acids as a monomeric building blocks.

As in most polyester syntheses, the challenge would be to reach high molecular weights. Various methods can be applied to polymerize HA to polyesters by condensation. Crucial features of these reactions are the activation of the carboxylic group and the removal of water to shift the equilibrium to the product side. HA can be activated by dicyclohexylcarbodiimide (DCC), *p*-toluenesulfonyl chloride (TosCl), or 2,4,6-triisopropylbenzenesulfonyl chloride (TPS) in anhydrous solvents at 25°C to start the polymerization reaction.[151] Triethyl amine (NEt$_3$) already triggered polymerization of HA at 0°C.[76] Boiling HCl or concentrated H$_2$SO$_4$ can also be used to catalyze the release of water.[152] For conversion via titanium(IV)isopropoxide temperatures of 140°C and reduced pressure are nessecary.[153] Seebach et al. obtained PHB from its monomers by adding (COCl)$_2$ and pyridine at -78°C.[154, 155] So far, only racemic starting materials or (*R*)-3-hydroxybutyric acid have been used for the previously described reactions.

Metabolism of PHA accumulation and degradation

In order to efficiently exploit the potentials of PHA, understanding of the genetic, biochemical and physiological basis for metabolism of PHA is of crucial importance. Although PHA production has been studied extensively, only the metabolism of PHB has been elucidated in greater detail.

Intracellular PHA granules

In vivo, PHA is formed in intracellular granules, having a core of amorphous PHA and on the surface a monolayer consisting mainly of specific proteins and phospholipids. Besides PHA polymerase and PHA depolymerase proteins, the main components of the surface layer are amphiphilic proteins named phasins with 10 to 40 kDa (e.g., PhaI 18 kDa and PhaF 36 kDa in *P. putida*).[37] They all contain a hydrophobic region, which is believed to anchor them to PHA; hence phasins can provide an amphiphilic layer between hydrophobic PHA and hydrophilic cytoplasm. In the absence of phasins, PHA storage is still possible, however, less PHA is accumulated and PHA granules are bigger, less numerous, and tend to coalesce.[156] Hence, phasins do not have an essential function in PHA metabolism; nevertheless, they are important for granule morphology and they might affect functionality of PHA processing enzymes.

Substrate specificity of the PHA polymerase determines which type of (co-)polymer can be obtained *in vivo*. Layered PHA granules with a core of mcl-PHA and a shell of poly(3-hydroxy-5-phenylvaleric acid) were observed in *P. putida* grown on 5-phenylvaleric acid plus nonanoic acid as carbon source.[157] Therefore, the experiment indicated a rather low substrate specificity of PHA polymerases.

If two different PHA polymerases with non-overlapping substrate range were present in a cell, the two types of PHA would be found in separate granules. Initiation of micelle formation seems to depend on the distinct type of PHA polymerase and only the same type of PHA polymerase is able to bind to a nascent granule.[33]

Physiological role of PHA

PHA plays an important physiological role for survival of bacteria cells in changing environments. Increased cell survival of PHA-containing cells was observed in carbon-depleted cultures of *Bacillus megaterium, Alcaligenes eutrophus*,[158] *P. putida* GPo1,[159] and *P. aeruginosa*,[160] when compared to their PHA-negative mutants.[161] Enhanced stress resistance towards heat or ethanol was also detected.[159] For example, a higher proportion of cells was vegetative in PHB-positive strains, whereas cells devoid of intracellularly stored carbon/energy reserves such as PHB earlier committed to sporulation.[158]

Enhanced survival and stress resistance can be explained by several modes of action of PHA: Under starvation conditions, it serves as carbon and energy source.[162-164] In addition, it can be used as electron donor,[165] i.e., as deposit for reducing equivalent.[42] For example, the oxygen concentration in the bacterial environment regulates the PHA content in nitrogen-fixing soil bacteria.[21, 166] In *C. necator*, PHA utilization was associated with respiration and oxidative phosphorylation and inhibited degradation of nitrogenous cellular constituents.[21] PHA degradation was also reported to be associated with nucleotide accumulation.[159] The products of polymer degradation can be used for the synthesis of nucleotides and other compounds. Higher levels of ATP and guanosine tetraphosphate (ppGpp) in turn induce expression of the *rpoS* gene, an indicator of cellular stress.

The transcription factor *rpoS* activates the expression of genes involved in stress response and in cellular shape change. The stress-induced rounder cell shape favors the absorption of nutrients from the environment.[159] Hence, *rpoS* activation by ppGpp is linked to PHA degradation and contributes to survival and stress tolerance.

In *P. aeroginosa*, PHA accumulation not only had a positive effect on stress tolerance but also influenced biofilm formation.[160] PHA-negative mutants formed a more stable and differentiated biofilm than corresponding wild type strains. Biofilm formation is associated with rhamnolipid and alginate biosynthesis. Since precursors for both PHA and rhamnolipid biosynthesis are derived from fatty acid *de novo* synthesis, and acetyl-CoA is precursor for both PHA and alginate, these reaction pathways are in competition. Absence of PHA polymerases directed carbon flux to alginate synthesis, i.e., biofilm formation. The stress tolerance of PHA-negative mutants was impaired. The regulation towards alginate synthesis is advantageous to nutrient-starved cells, because biofilm formation offers protection against environmental stresses.[160]

Metabolic pathways for PHA

PHA are accumulated if a carbon source is provided in excess and at least one other nutrient, essential for growth, is depleted.[21] PHA polymerases are the key enzymes that polymerize hydroxyacyl-CoA thioesters to form PHA. Hydroxyacyl-CoA thioesters are the precursor molecules for PHA. They are either obtained from central intermediates of carbon metabolism or from derivatives of an available carbon source.

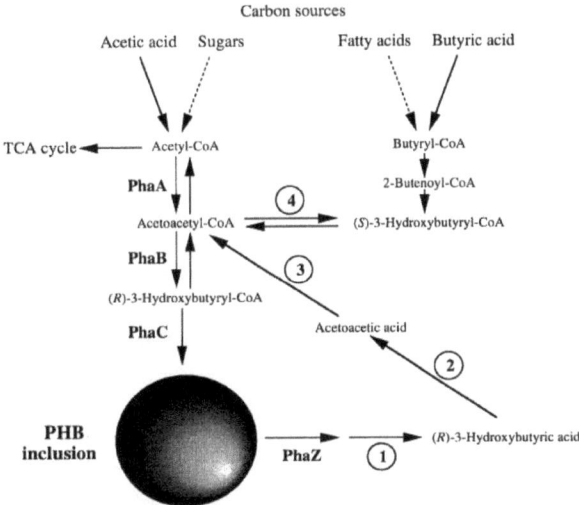

Figure 1.3 Cyclic metabolic nature of PHB biosynthesis and degradation in bacteria. PhaA, β-ketothiolase; PhaB, NADPH-dependent acetoacetyl-CoA reductase; PhaC, PHA polymerase; PhaZ, PHA depolymerase; 1, dimer hydrolase; 2, (R)-3-hydroxybutyrate dehydrogenase; 3, acetoacetyl-CoA synthetase; 4, NADH-dependent acetoacetyl-CoA reductase (from Sudesh et al.[78]).

For PHB, 3-hydroxybutyryl-CoA can naturally be provided by two pathways (Fig. 1.3). In most bacteria, two acetyl-CoA molecules are converted to acetoacetyl-CoA by a Claisen-condensation catalyzed by a β-ketothiolase, and then reduced to (R)-3-hydroxybutyryl-CoA mostly by a NADPH-dependent acetoacetyl-CoA reductase.[167] Another pathway was found in *Rhodospirillum rubrum*, where acetoacetyl-CoA is first reduced to (S)-3-hydroxybutyryl-CoA and subsequently converted to (R)-3-hydroxybutyryl-CoA by a stereospecific enoyl-CoA hydratase.[168]

Other scl-PHA are formed when corresponding precursor substrates are provided. Even compounds such as α,ω-alkanediols,[169] lactones,[170] or ketoacids[171] can serve as PHA precursors, being processed by β-oxidation prior to incorporation. Reaction pathways of scl-PHA are well investigated, e.g., metabolic engineering for overproduction as well as the large scale synthesis of PHBV with technically suitable properties[172] or PHA production in transgenic plants[173] has been shown.

Up to now, a cyclic reaction mechanism for PHB metabolism has been commonly accepted. A metabolic route for PHB biosynthesis and degradation (Fig. 1.3) was proposed.[42, 44] Simultaneous synthesis and degradation of PHB was confirmed by ^{14}C-glucose pulse experiments.[39] According to de Roo,[92] the metabolic advantage of such a mechanism is doubtful as a significant amount of energy is wasted. But he suggests that this cycle provides the cell with a valuable regulatory mechanism to control fatty acid flux through the fatty acid pathway.[92] The concentrations of intracellular metabolites determine the direction of conversions within the reaction cycle. With respect to physiological considerations, PHA might function as a readily available energy and carbon reservoir.[15, 166, 174, 175] Its function as an electronic sink into which reducing power can be channeled has also been discussed.[176, 177]

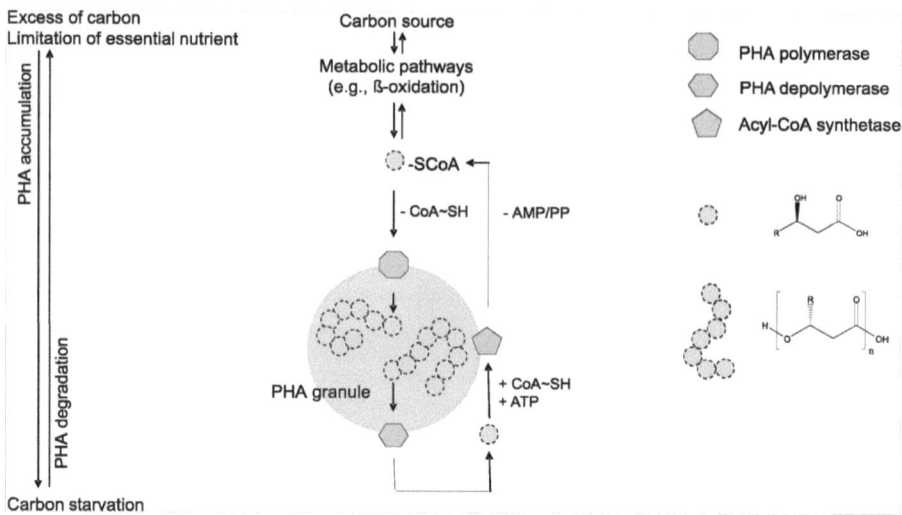

Figure 1.4 Scheme of mcl-PHA metabolism. Reactions leading to PHA accumulation are favored with excess of carbon and limitation of essential nutrients. Net PHA degradation occurs under carbon starvation. Enzymes involved are discussed below.

The metabolism of mcl-PHA has not yet been described to the last detail (Fig. 1.4). Mcl-PHA was first discovered in *P. putida* GPo1 grown on alkanes.[178] For example, feeding *n*-octane resulted in poly((*R*)-3-hydroxyoctanoate-*co*-(*R*)-3-hydroxyhexanoate) (PHO). Amongst others, the production of mcl-PHA is for instance a typical feature of fluorescent pseudomonads.[179]

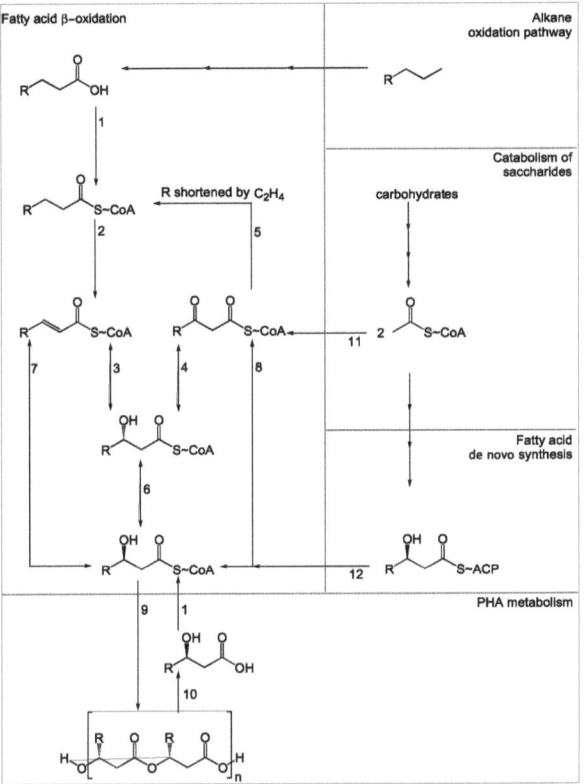

Figure 1.5 Metabolic pathways of *P. putida* showing origin of PHA precursors (fatty acid β-oxidation, fatty acid *de novo* synthesis, alkane oxidation pathway or catabolism of saccharides) and PHA metabolism. Reaction conditions are the following (adapted from Kunau et al.[180]): 1: acyl-CoA synthetase (+HS~CoA, +ATP, -AMP/PP, -H$_2$0); 2: acyl-CoA dehydrogenase (+FAD, -FADH$_2$); 3: enoyl-CoA hydratase (+H$_2$0); 4: NAD-dependent (*S*)-3-hydroxyacyl-CoA dehydrogenase (+NAD, -NADH$_2$); 5: 3-ketothiolase (+HS~CoA, -acetyl-S~CoA); 6: 3-hydroxyacyl-CoA epimerase; 7: hypothetical R-specific enoyl-CoA hydratase (-H$_2$0); 8: hypothetical NADPH-dependent 3-ketoacyl-CoA reductase (+NADP, -NADPH$_2$); 9: PHA polymerase (-HS~CoA); 10: PHA depolymerase (+H$_2$0); 11: β-ketothiolase (-HS~CoA); 12: 3-hydroxyacyl-ACP-CoA transferase (+CoA, -ACP).

Two major pathways lead to (*R*)-3-hydroxyacyl-CoA, i.e., the precursors of PHA; which of the two is more relevant depends on the organism. When only non-related carbon sources such as carbohydrates are available, *de novo* fatty acid synthesis can be involved (Fig. 1.5). When cells grow on fatty acids or the corresponding alkanes, monomer synthesis occurs through the β-oxidation pathway. In the first step, alkanes are oxidized to fatty acids, which are then subsequently degraded via fatty acid β-oxidation pathways. The link to provide 3-hydroxyacyl-CoA for PHA polymerase occurs via either an enoyl-CoA hydratase (using *trans*-2-enoyl-CoA as substrate),[181, 182] or a ketoacyl-CoA reductase (using ketoacyl-CoA as substrate), or an epimerase converting *S*-3-hydroxyacyl-CoA into *R*-3-hydroxyacyl-CoA.[183, 184] The mcl-PHA metabolism might follow a cyclic reaction similar to PHB metabolism (Fig. 1.5).

A major metabolic difference between strains is whether or not intermediates from *de novo* fatty acid synthesis are incorporated into the polymer. The two closely related strains *P. putida* GPo1 and *P. putida* KT2442 have been investigated with respect to their ability to accumulate PHA:[185] GPo1 only produces PHA under nitrogen limitation when structurally similar carbon compounds are available (i.e., fatty acids). In contrast to this, KT2442 is also able to build up PHA when grown on non-related carbon substrates (e.g., glucose) even without nitrogen limitation. This is due to a transferase in KT2442 which can channel suitable intermediates of the *de novo* fatty acid pathway to the PHA polymerases. This enzyme is not functional in GPo1, thus regulation of PHA metabolism is different in the two strains.

In fact, many pseudomonades, excluding *P. putida* GPo1, can accumulate PHA from simple, non-related carbon sources such as glucose.[186, 187] The major pathway to provide 3-hydroxyacyl-CoA is the fatty acid *de novo* synthesis[188] and a transferase channeling 3-hydroxyacyl from the acyl carrier protein (ACP) into its CoA activated form (e.g., catalyzed by PhaG$_{Pp}$ in *P. putida*).[185] The resulting polyesters mainly consist of 3-hydroxydecanoate with small amounts of 3-hydroxydodecanoate and 3-hydroxyoctanoate.[186, 188] In recombinant *E. coli* harbouring the corresponding genes from the PHA operon overproduction of various scl-PHA[189] and synthesis of mcl-PHA[190-192] has been demonstrated.

Products of PHA degradation can be monomers, dimers, oligomers, or mixtures thereof.[193, 194] Reaction pathways and involved hydrolases have been analyzed to some extent for scl-PHA,[195-198] but for mcl-PHA the reaction mechanisms of intracellular PHA decomposition still need to be further clarified. In order to understand the regulatory mechanisms, research on enzymes involved, e.g., the almost unknown role of acyl-CoA synthetases is needed for a deeper insight into the metabolic pathways of PHA accumulation and degradation. The most important enzymes are described in the following sections. So far, investigations concerning underlying principles of PHA metabolism and its genetics have contributed a lot to efficient synthesis of tailor-made PHA and related products.

PHA polymerases

PHA polymerases are constitutively expressed enzymes that are localized at the surface of PHA granules[199, 200] and that are soluble in the cytoplasm in non-PHA-accumulating cells.[77] PHA polymerases catalyze the intermolecular ester bond formation between the hydroxyl and the carboxyl group starting from CoA thioesters of the corresponding HA. The enzymes can be classified into four different types (I, II, III and IV) with respect to their size, structure, and substrate specificity.[33] Type I polymerases have one subunit and occur in most heterotrophic bacteria and phototrophic nonsulfur purple bacteria. The 64 kDa PhaC$_{Re}$ of *C. necator* is representative for this type. Type II polymerases also have one type of subunit and occur in all pseudomonades belonging to the rRNA homology group I. The 62 kDa PhaC1$_{Pa}$ of *P. aeruginosa* is a representative of this type II. Type III consists of two different subunits and is represented by, e.g., PhaC$_{Cv}$ (40 kDa) and PhaE$_{Cv}$ (41 kDa) of *Chromatium vinosum*.[201] The active enzyme is probably a decamer with PhaC$_{Cv}$ being the catalytically active subunit. A type IV PHA polymerase has recently been found in *Bacillus megaterium* that contains two subunits (PhaC 40 kDa and PhaR 22kD), and produces scl-PHA.[202]

The substrate specificity of PHA synthetases is extremely wide with respect to the length of the alkyl residue, the presence of additional functional groups or the position of double bonds. That is why so many different monomers are found to be incorporated into PHA. However, stereo specificity is strict, demanding the hydroxyl carbon atom to have *R*-configuration. Type I and III polymerases build up scl-PHA and the hydroxyl group can be at position 3, 4, or 5. Type IV seems to accept only 3-hydroxyl compounds for scl-PHA. Type II polymerases preferentially start with 3-hydroxyl compounds to catalyze the formation of mcl-PHA. Exceptions of this classification are for example PHA polymerases of *Thiocapsa pfennigii* and *Aeromonas caviae*, which catalyze the polymerisation of co-polyesters of scl- and mcl-PHA.[203-206]

The reaction mechanism of PHA synthesis can be explained by the following cycle, valid for all PHA polymerases so far: The active centre is a highly conserved cystine residue, which binds a CoA-activated hydroxycarboxylic acid (1) upon release of HS-CoA. A second hydroxycarboxylic acid (2) is bound to a second active centre. Then, the hydroxyl group of the acyl thioesters (1) adds to the carbonyl group of acyl thioesters (2) via nucleophilic attack, and subsequently the thioesters bond of the acyl thioesters is eliminated. This active centre is now free to accept a new CoA activated hydroxycarboxylic acid. The catalytic cycle is repeated several thousand-times resulting in high molecular weight polyesters. The reaction is terminated by the attack of other nucleophils such as water or glycerol. The second active centre might be provided by a second cys-residue,[207] or a conserved serine residue,[208] or by the same conserved cys of a second subunit.[209] So far, the total number of subunits present in active PHA polymerase has not been elucidated. When scl-PHA and mcl-PHA polymerase are co-expressed, the recombinant bacteria produce a blend of two different polyesters and not the corresponding copolyester,[210] which proves that the growing PHA chain remains covalently bound to the enzyme. CoA inhibits the PHA polymerases of *C. necator*, *C. vinosum* and *P. aeruginosa* at low concentrations.[52, 211]

PHA depolymerases

Intracellular PHA depolymerases

Intracellular PHA degradation is an active mobilization of endogenous PHA by the accumulating bacterium itself, which is triggered under carbon starvation and which is accomplished by intracellular PHA depolymerases.[15] These enzymes can degrade amorphous PHA in native granules. Even though, decomposition of PHA has been demonstrated in many bacteria long time ago,[212-215] mechanisms and regulation of the intracellular degradation of PHA are not completely understood.[15] Depolymerase activities depend very much on the properties of the PHA polymer; for example, intracellular PHA depolymerases are not able to hydrolyze crystalline PHA. Amorphous PHA for enzyme assays can be obtained for example by purifying PHA granules in the native form. When incubated under appropriate conditions, auto-hydrolysis occurs due to granule-associated proteins.[15, 216-218] For example, native poly(3-hydroxy octanoate) granules of *P. putida* showed self-hydrolysis with an optimum at pH 9.[216-218]

The intracellular PHB depolymerase of *R. rubrum* was purified. It is a 35 kDa polypeptide structurally related to α/β-proteins with a type II catalytic domain, having a catalytic triad and a lipase box (Gly-Xaa_1-Ser-Xaa_2-Gly). The active centre is the catalytic triad with the three conserved amino acids serine, aspartate, and histidine. The N-terminus shows a signal peptide sequence, which might direct the protein to the periplasm.[219] In *C. necator*, an intracellular depolymerase was found that did not contain any lipase-box fingerprint. It did not show similarities to the enzyme of *R. rubrum*, and had two pH optima at pH 7 and 9.[220, 221] Nine potential PHB depolymerase/oligohydrolase genes have been identified in *C. necator*.[35, 222, 223] The characterization of some other PHB depolymerase genes was also successfully accomplished,[193, 220] however, more research is needed to get a detailed picture.

Degradation of mcl-PHA has not been elucidated in great detail so far. A mcl-PHA depolymerase activity has been shown in *P. putida* GPo1 and U by mutagenesis and consecutive complementation experiments.[38, 224] Depolymerases with potential lipase box pentapeptides were identified in several pseudomonades and related organisms, but no significant homologies to any scl-PHA depolymerase were detected. Recently, the first intracellular mcl-PHA depolymerase, *PhaZ* from *P. putida* KT2442, was purified and characterized.[46] It is located on the PHA granule surface and hydrolyzes specifically mcl-PHA containing aliphatic and aromatic monomers. Primarily reaction products are the corresponding dimers and ensuing free 3-hydroxycarboxylic acids are formed as main product. Hence, *PhaZ* behaves as *endo-* and *exo-*hydrolase. Its optimal activity is at pH 8.8 and 43°C in buffer with high ionic strength (300mM NaCl). The enzyme is classified as a serine hydrolase being inhibited by phenylmethylsulfonyl chloride. Modelling revealed an α/β-hydrolase-type domain capped with a lid structure and a catalytic triad as active site. The presence of a lid domain in *PhaZ* from *P. putida* KT2442 is the main difference to extracellular mcl-PHA depolymerase from *P. fluorescens* GK13, and might explain why the first one has a narrower substrate specificity range whereas the latter can also hydrolyze, e.g., water-soluble esters. Sequence comparisons in *P. putida* KT2442 revealed at least 14 genes similar to *phaZ*. Their functions are not yet determined but their products might be involved in a more elaborate system of PHA mobilization.[46] Further putative depolymerase genes for intracellular mcl-PHA have been discovered in other organisms but structural and functional data are still very rare.

Extracellular PHA depolymerases

In contrast to intracellular PHA degradation, extracellular degradation is carried out by microorganisms that secrete extracellular depolymerases, which utilize exogenous polymers, e.g., PHA released from cells after death and cell lysis.[15] The products can be utilized as carbon and energy source by bacteria and fungi.[48, 225] Extracellular PHA degradation requires different degradative enzymes than intracellular PHA degradation because PHA exhibit different biophysical states *in vivo* (native/amorphous)[226] and outside of cells (denatured/semi-crystalline).[227] Microorganisms able to degrade extracellular PHA are widespread in the environment; they have been isolated from various ecosystems.[48] Over twenty bacterial extracellular PHA depolymerase genes have been cloned[228, 229] and several depolymerases from bacteria and fungi have been investigated,[48] e.g., in particular the extracellular mcl-PHA depolymerase from *P. fluorescens* GK13.[229] These enzymes mostly consist of one <70 kDa polypeptide and show a high stability at a wide range of different pH values with an activity optimum at alkaline pH. Many extracellular scl-PHA depolymerases are inhibited by reducing agents and by serine hydrolase inhibitors.[15] Molecular biological analyses revealed the following general domain structure for extracellular PHB depolymerases:[15] A signal peptide for secretion is aligned prior to a catalytic domain at the N-terminus of the mature protein. An active centre with a catalytic triad occurs as well as a lipase-box. Two types of catalytic domain have been identified, differing in their arrangement of the catalytic active amino acids. A linking domain connects the catalytic domain with a substrate-binding domain (SBD) at the C-terminus. The SBD binds to PHB with some specificity for its chemical and physical properties, especially regarding its crystallinity and side chain length. Some conserved amino acids have been identified at the SBD but it is not known whether they are directly involved in binding of the polymer chain.[15, 230]

The ability to degrade PHA is widely distributed among bacteria and fungi, and the underlying mechanisms have been a field of intensive research. The synthesis of extracellular PHA depolymerases in bacteria is generally repressed if suitable soluble carbon sources such as glucose or organic acids are present.[15] After exhaustion of nutrients, synthesis of PHA depolymerases is derepressed in many strains.[231] Even though still some questions have to be clarified, in general, much more is known about extracellular PHA depolymerases than the intracellular ones.

Acyl-CoA synthetases

In order to recycle PHA monomer acids, cells have to activate them to a CoA- or ACP-linked form by enzymes such as acyl-CoA synthetase. These enzymes might convert the HA released by PHA depolymerase PhaZ and, in this way, render the monomers either accessible for β-oxidation or make them reusable for PHA production. Since the diffusion of monomer acids away from the PHA granule into the cytoplasm may lead to a cytoplasmic pH change, it would be more efficient and practical if the enzyme for activating the monomer acid was also located on the PHA granule surface.

Acyl-CoA synthetases (ACS) are a ubiquitous family of enzymes that activate fatty acids by ligating CoA to their carboxy residues.[232-234] The activation of acetic acid and fatty acids is conserved in nature.[234] The reaction mechanism involves two steps: 1. Acetate activation to acetyl-AMP; 2. Conversion of acetyl-AMP to acetyl-CoA.[234] The products are substrates for acyl-CoA dehydrogenases in the first conversion of the β-oxidation pathway. Most ACS are involved in the β-oxidation at the fatty acid degradation pathways.[235-237] They all show two highly conserved regions, one of which is the typical glycine-rich loop (YTSG(S/T)TGxPKG) of an ATP-binding domain.[238]

One of the most extensively studied ACS is FadD of *E. coli*.[239, 240] FadD is a cytoplasmic membrane-associated protein, and activates exogenous long-chain fatty acids into metabolically active CoA thioesters when they are transported across the cytoplasmic membrane.[239, 241] The *E. coli* genome contains one ACS-encoding gene, *fadD*, whereas most *P. putida* strains hold two homologous genes (Fig. 1.6).

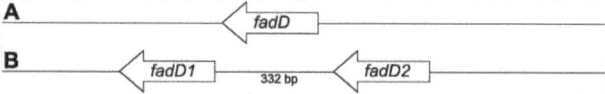

Figure 1.6 (A) *E. coli* K12 contains one acyl-CoA-synthetase (ACS) encoded by *fadD* (1685 bp). (B) *P. putida* KT2440 contains two ACS encoded by *fadD1* (1697 bp) and fadD2 (1688 bp).[242, 243]

In *P. putida* U, the presence of two FadD homologues has been reported: FadD1 and FadD2.[38] One gene product is a homodimer (141 kDa) which converts *n*-alkanoic acids (C_2-C_8) as well as some aromatic compounds (phenylacetic and phenoxyacetic acid)[244] and shows higher activity for shorter chain length HAs. The other acyl-CoA-activating enzyme in *P. putida* U is not activated by acetate, however, once it is induced, it is also capable to convert acetate.[245] It was reported that this enzyme was a phenyl-acetyl-CoA ligase[246] inducible by phenylacetic acid and with a broad substrate specificity. The *fadD1*-deficient-mutant is not able to grow on fatty acids with acyl chains longer than C_4 as sole carbon sources.[247, 248] However, prolonged incubation of 80 hours could restore growth.[248] Disruption of the *fadD2* gene did not have any effect on the catabolism of fatty acids.[248] It was concluded that FadD1 is an enzyme involved in the physiological degradation of fatty acids, and FadD2 is induced when FadD1 is inactivated. Once adapted, *fadD1* mutants of *P. putida* U were able to resume growth and PHA synthesis, reaching similar biomass and PHA contents as the parental strain.[248] The produced PHA has similar morphology and monomer composition as the parental strain.[248] Hence, there is evidence that acyl-CoA synthetases are involved in activating PHA monomers to their thioesters, however, not much is known about the role of this enzyme in PHA metabolism.

Aim and scope of this thesis

The aim of this project is to elucidate PHA degradation in *P. putida* GPo1 and, based on the findings, establish a production method for PHA monomers (*R*-3-OH-carboxylic acids, HA). Disadvantages of current production methods of HA are low yields, low enantiomeric excess and complicated, often costly, strategies, which mostly lack the possibility to recycle solvents. Furthermore, degradation mechanisms of mcl-PHA are not yet fully understood. Physiological conditions that initiate PHA degradation and enzymes involved in the conversion of degradative products require further investigation. These questions will be addressed in the following chapters. On the one hand, a practical and efficient approach to isolate different HA will be investigated to supply them as fine chemicals for industrial applications. On the other hand, basic research on the degradation pathways in *P. putida* GPo1 will be carried out. The intention is to investigate the underlying principles and understand the degradation process on a molecular level.

Previously, different approaches to obtain HA have been used, including organic synthesis as well as chemical hydrolysis of purified PHA.[80] These approaches either give low yields or low enantiomeric excess, or consume large amounts of solvents. Therefore, in this research a direct *in vivo* depolymerization method will be investigated (**Chapter 2**). To achieve *in vivo* depolymerization, appropriate physiological conditions have to be found and optimized where PHA depolymerase is still active, whereas cells cannot further utilize the produced monomers. This would provide a quicker, cheaper, and less complex approach compared to what was reported before.[70, 249] A method will be developed to integrate the new approach into a new product manufacturing process. The optimization of conditions for chiral synthon production should result in an economic, easy-to-handle, fast, and environmentally friendly process. PHA accumulation can be accomplished by cultivation of *P. putida* GPo1 in continuous culture. Monomer release can be achieved by *in vivo* depolymerization. The analytical measurements of PHA and HA compositions will be carried out by high performance liquid chromatography mass spectroscopy (HPLC-MS) and by gas chromatography (GC), applying an improved method for HA quantification.[250] Different synthons can be produced using different carbon feeds during cultivation, which will lead to PHA polymers with various monomer compositions. For interesting synthons, production methods for larger scales will be designed.

For future industrial applications, it is not only important to be able to produce HA but it is even more relevant to isolate them at a high degree of purity (**Chapter 3**). Microbiologically produced PHA is often a co-polymer. This is a result of the fatty acid β-oxidation which removes C_2 units from PHA precursors.[36] Monomers obtained from PHA of *P. putida* GPo1 consequently are a mixture of HA whose side chains differ in the length of ethylene subunits (C_{n-2i}). They are chemically very similar and therefore difficult to separate. Thus, it is necessary to set-up experiments for separation and purification of these synthons produced from bacterial PHA. Solvent extraction and column chromatography methods will be tested and the purification process can be monitored by HPLC-MS and other analytical methods. The developed process should be environmentally friendly, with the possibility of solvent recycling and should have the potential to be scaled up for industrial applications.

In order to further improve HA production, a continuous *in vivo* bioprocess will be established by exploiting both the PHA-synthesizing and PHA-degrading abilities of *P. putida* GPo1 (**Chapter 4**). The setup includes a chemostat for controlled PHA accumulation and an interconnected second reactor where pH is shifted to the alkaline range. These conditions are designed to force HA release to the medium. In combination with high cell density cultivation this novel bioprocess will provide an economic and environmentally friendly way to produce chiral compounds needed for the synthesis of pharmaceuticals, vitamins and flavours. A patent application for this procedure is aspired.

Accumulation and degradation of PHA is a physiological advantage for cells concerning survival and stress tolerance.[159] Here, the physiological advantage of excreting HA to the media for *P. putida* GPo1 will be investigated (**Chapter 5**). Since *in vivo* depolymerization in *P. putida* GPo1 was only observed under alkaline conditions, it can be speculated that cells release HA to neutralize their environment and thus enhance their survival rate under harsh alkaline conditions. A better knowledge of the physiological situation during PHA degradation is thought to be helpful to optimize PHA monomer production.

Chapter 1

Measurements of cell viability, of extracellular pH and of intracellular pH using fluorescence-labelling[251] will be carried out in order to elucidate physiological conditions that might trigger PHA degradation in the bacteria.

Within the scope of this project, an acyl-CoA synthetase (ACS) potentially involved in PHA degradation is investigated, including its activity and cellular location (**Chapter 6**). The ACS is proposed to activate PHA monomers, which can be further utilized by cells as carbon and energy source (see Fig. 1.4). Measuring the enzymatic activity of an acyl-CoA synthetase was described by de Roo.[92] He isolated PHA granules from cells by density centrifugation and observed depletion of CoA, thus indicating activity of an ACS. Alternatively, the formation of the (R)-3-hydroxycarboxylic thioesters can be monitored by HPLC to measure ACS activity. In order to explore the regulation of PHA degradation at an enzymatic level, activity tests of ACS on different substrates will be accomplished. *In vivo* localization of two ACS found in *P. putida* GPo1 will be tested by construction of gfp-fusions and investigation of the obtained mutants by fluorescence microscopy under various conditions.

Based on previous observations, we presumed the existence of an ACS which is directly involved in PHA degradation in *P. putida* GPo1. In order to further elucidate the function of this enzyme, a knock-out mutant will be constructed (**Chapter 7**). In theory, knocking out the corresponding *acs* should block the metabolic re-utilization of monomers (compare Fig. 1.4). ACS can be disrupted by means of insertion of a kanamycine resistance gene. The question will be addressed whether or not this gene knockout would block degradation of PHA monomers and hence PHA accumulation would increase. We suggested that higher concentrations of PHA and PHA monomer might be reached by cultivation of this ACS-negative mutant, since the degradation PHA monomers would be interrupted.

Chapter 2

Bacterial poly(hydroxyalkanoates) as a source of chiral hydroxyalkanoic acids

Qun Ren, Andreas Grubelnik, Mirjam Hoerler, Katinka Ruth, René Hartmann, Helene Felber, and Manfred Zinn. 2005. Biomacromolecules, 6 (4), 2290-2298

Katinka Ruth has carried out the following parts: Proof of concept: effect of pH, temperature and time on *in vivo* depolymerization of PHA, qualitative and quantitative determination of PHA monomers.

Abstract

Polyhydroxyalkanoates (PHA) are polyesters of various hydroxyalkanoates accumulated in numerous bacteria. All of the monomeric units of PHA are enantiomerically pure and in R-configuration. R-hydroxyalkanoic acids can be widely used as chiral starting materials in fine chemical, pharmaceutical and medical industries. In this study we established an efficient method for the production of chiral hydroxyalkanoic acid monomers from PHA. *Pseudomonas putida* cells containing PHA were re-suspended in phosphate buffer at different pH. We observed that the optimal initial pH for intracellular PHA degradation and monomer release was at pH 8-11 with pH 11 as the best. At initial pH 11, PHA containing 3-hydroxyoctanoic acid and 3-hydroxyhexanoic acid was degraded with an efficiency of over 90% (w/w) in 9 hours, and the yield of the corresponding monomers was also over 90%. Under the same conditions unsaturated monomers were also effectively produced from PHA containing 3-hydroxy-6-heptenoic acid, 3-hydroxy-8-nonenoic acid and 3-hydroxy-10-undecenoic acid. The monomers (e.g., 3-hydroxyoctanoic acid) were further isolated using solid phase extraction and purified on reversed phase semi-preparative liquid chromatography. We confirmed that the purified 3-hydroxyoctanoic acid monomer has exclusively R-configuration.

Introduction

Polyhydroxyalkanoates (PHA) are naturally occurring polyesters that are produced by a wide variety of microorganisms.[13] They can be classified into three groups based on the number of carbon atoms in the monomer units:[13] short-chain-length (scl) PHA, which consists of 3-5 carbon atoms, medium-chain-length (mcl) PHA, which consists of 6-14 carbon atoms, and long-chain-length (lcl) PHA, which consists of more than 14 carbon atoms. PHA is normally synthesized when cells are cultured in the presence of an excess carbon source and when growth is limited by the lack of an essential nutrient.[252] If the cells are under carbon limitation, the accumulated PHA can be degraded to the monomers and can be re-utilized by the bacteria as a carbon and energy source (Fig. 2.1).[252] PHA is suitable for a broad range of applications in medicine, pharmacy and industry due to its biocompatibility and biodegradability.[14, 69, 167] Furthermore, all of the PHA monomers are enantiomerically pure and in R-configuration.[79] More than 140 types of 3-, 4-, 5- or 6-hydroxyalkanoic acids have been found to be incorporated into the polymer,[13, 78] and an increasing number of new monomers are being discovered. R-3-hydroxyalkanoic acids are valuable synthons and can be widely used as starting materials for synthesis of antibiotics, vitamins, flavors and pheromones.[80, 81] Thus, PHA displays a unique potential as a source of R-3-hydroxyalkanoic acids and derivatives thereof.

Different processes have been reported for the preparation of enantiomerically pure 3-hydroxyalkanoic acids. Enantiopure 3-hydroxyesters have been obtained by using 3-ketoesters or 3-ketoacids as prochiral precursors.[84, 253] However, this process suffers from several limitations, such as pure substrates are required, and the range of possible products is limited. The alternative process of biocatalytic reduction of β-keto esters using bakers' yeast or other microorganisms avoids these problems.[88, 91] However, this biocatalytic approach often gives moderate yields and a relatively low enantiomeric excess. Moreover, it requires expensive starting compounds or complex multi-step reactions which entail high production costs. More recently, methods for producing R-3-hydroxyalkanoic acids by chemical digestion of PHA have also been reported.[92, 93] However, large amounts of organic solvents were used and the production efficiency was rather low due to multi-step processes.

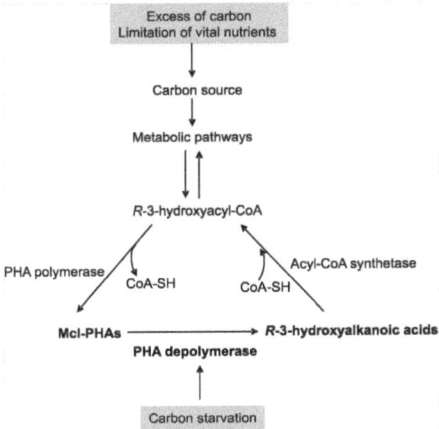

Figure 2.1 Metabolism of mcl-PHA. A metabolic cycle is formed by the sequential action of the PHA polymerase PhaC, PHA depolymerase PhaZ, and acyl-CoA synthetase (adapted from Witholt and Kessler[252]). When the carbon source is in excess and other vital nutrients are limited, PHA accumulation will occur, whereas when carbon source is limited, PHA degradation will take place. The boxed compounds illustrate the effectors that stimulate the direction of the pathway.

An attractive alternative to obtain R-3-hydroxyalkanoic acids is the *in vivo* depolymerization of PHA. The key enzymes for mobilization of PHA are intracellularly located PHA depolymerases, which are able to hydrolyze PHA.[15, 94, 95] The produced 3-hydroxyalkanoic acid monomers are excreted into the medium and can be further isolated. Previously, Lee and co-workers demonstrated that R-3-hydroxybutyric acid (R3HB) could be efficiently produced via *in vivo* depolymerization in naturally poly(3-hydroxybutyrate)-producing bacteria by providing appropriate environmental conditions.[70] In their study with the strain *Alcaligenes latus*, they found that lowering the pH to 3-4 induced the highest activity of intracellular poly(3-hydroxybutyrate) (PHB) depolymerase and blocked the re-utilization of R3HB by the cells.[70] The authors investigated the *in vivo* depolymerization of mcl-PHA in *Pseudomonas* as well, however, the production efficiency of mcl-3-hydroxyalkanoic acids was very low (maximum 9.7% after 4 days).[70]

The metabolism of mcl-PHA has different pathways from that of the well studied scl-PHA.[252] For *P. putida* GPo1 (previously known as *P. oleovorans*), the genes involved in PHA biosynthesis and degradation have been cloned and sequenced.[17] A depolymerase gene *phaZ* was identified between two polymerase-encoding genes *phaC1* and *phaC2*.[17] This depolymerase has been reported to be a PHA granule associated, intracellular protein.[15, 94, 95] The intracellular depolymerases have a general range of pH optimum at 8-10.[15] At present, the mechanism and regulation of the intracellular degradation of mcl-PHA (i.e., the mobilization of previously accumulated PHA) are poorly understood.[15]

This paper describes a simple and effective method to produce mcl-3-hydroxyalkanoic acids via *in vivo* depolymerization of mcl-PHA. Based on this method mcl-3-hydroxyalkanoic acids could be produced in *P. putida* GPo1 with a yield of over 90% (w/w). One monomer (3-hydroxyoctanoic acid) was further purified and the absolute configuration was determined by chiral GC analysis. The reported method provides an economical system to obtain enantiomerically pure 3-hydroxyalkanoic acids, which might be the first step towards a large scale production of chiral synthons in industry.

Experimental Section

Bacterial strains and cultivation conditions

Wild-type strains *P. putida* GPo1 and *P. putida* U, and their depolymerase-deficient mutants *P. putida* GPo500[254] and *P. putida* U phaZ[-38] were used in this study. In batch culture, cells were grown in modified medium E2[63] supplemented with 10 mM octanoic acid (Fluka, Switzerland) as carbon source and incubated in shaking flasks at 30 °C and 160 rpm.

The continuous cultivation was performed in continuous culture medium supplied with mineral trace element (CCMT).[31, 63] 40 L of this medium were filter sterilized into gamma sterilized 50 L medium bags (Flexboy, Stedim S.A., Aubagne Cedex, France). Carbon source (octanoic or 10-undecenoic acid) was pumped directly into the culture vessel by using a dosimat (Metrohm, Herisau, Switzerland). A 3.7 L laboratory bioreactor (KLF 2000, Bioengineering, Wald, Switzerland) with a working volume of 2.8 L was used. Cells grew on octanoic acid (Fluka, Switzerland) or 10-undecenoic acid (Fluka, Switzerland) in CCMT at a dilution rate of $D = 0.1\ h^{-1}$, and a carbon to nitrogen ratio (C/N) of 19 (g/g) or 15 (g/g) was applied. The cultures were run at 30°C, and the pH was maintained at 7.0 by automated addition of either 2 M NaOH or 2 M H_2SO_4. The dissolved oxygen tension was monitored continuously with an oxygen probe (Mettler Toledo, Greifensee, Switzerland) and care was taken that it remained above 35% air saturation. The culture volume was kept constant with an overflow tube that was connected to a continuously running harvest pump. The culture was collected in a 10 L harvest tank which was cooled on ice.

Conditions for preparation of 3-hydroxyalkanoic acids

Cells containing PHA were collected and resuspended in distilled water or 50 mM potassium phosphate buffer with different pH. The pH of the phosphate buffer was adjusted with 5 M KOH or 10 M HCl to the values of 2, 4, 7, 8, 9, 10, 11 or 12. The cell suspensions were incubated in sterile flasks at 30°C without shaking. The beginning of the incubation was set to be time-point zero. Samples were taken at selected time points to detect the monomer content by high performance liquid chromatography (HPLC) coupled with electrospray ionization mass spectrometry (ESI-MS).

Intracellular PHA analysis

The PHA hydrolytic step was adapted from the procedure for poly(3-hydroxybutyrate).[255] A known amount of about 10 mg lyophilized cells were weighed into a 10 mL pyrex tube. Then, 1 mL of methylene chloride containing 1.5 mg 3-hydroxyisovaleric acid as internal standard and 1 mL of a mixture of n-propanol / hydrochloric acid (80/20 v/v) were added. The tube was capped and heated for 3h at 100°C. After cooling, 2 mL of nano-pure water were added and the tube shaken on a laboratory mixer. The organic layer was then dried with anhydrous sodium sulfate. The derivatized samples were analyzed in methylene chloride solution on a GC (Hewlett Packard 5890/II, Urdorf, Switzerland) equipped with a flame ionization detector (FID). The separation was made on a Supelcowax 10 column, 30 m x 0.25 mm, 0.5 µm (Supelco, Buchs, Switzerland). The GC parameters were as follows: temperature of the injector 250°C, temperature of the FID detector 285°C, He gas flow 3 mL/min, split ratio 1:10, 3 µL of injection, and the oven temperature program was: 120°C, 1 min isotherm, 120-280°C with 10°C/min, 1 min isotherm. The propylesters of 3-hydroxyacid monomers were identified by comparing the sample retention times with the commercially available 3-hydroxy standards of butanoic, hexanoic, octanoic, decanoic, and dodecanoic acid after propanolysis. Quantification was done via a calibration function generated from a mixture of 3-hydroxyisovaleric acid and the 5 standards after propanolysis as described above, and interpolating the response factors for monomers not available commercially.

Analysis of 3-hydroxyalkanoic acid monomers

Samples were centrifuged at 20,000 x g and 4°C for 5 min. The supernatant was collected and diluted in 50% (v/v) acetic acid (0.1% (v/v) in water) and 50% (v/v) acetonitrile. The dilution factor was either 100 or 200 times depending on the signal intensity, i.e., the monomer content. For calibration we used different concentrations of 3-OH-octanoic acid (3-OH-C8) (Sigma-Aldrich, Seelze, Germany). For 3-OH-hexanoic acid (3-OH-C6), 3-OH-6-heptenoic acid (3-OH-C7:1), 3-OH-8-nonenoic acid (3-OH-C9:1) and 3-OH-10-undecenoic acid (3-OH-C11:1), there was no pure reference material available, thus it was not possible to quantify these monomers. Since the ionized groups ($-CHOH-CH_2-COO-$) of these compounds for HPLC-MS are identical, the signals of these compounds are comparable. Thus, we used the standard curve of 3-OH-C8 to obtain the relative content of other monomers based on the HPLC peak areas.

PHA monomers were separated on a C18 Nucleosil column (2 x 250 mm, 3 um, 100 Å, Macherey-Nagel Inc., Easton, PA, USA). Separation was achieved using a linear gradient from 0% to 100% acetonitrile (Sharlau, supragradient HPLC grade, Barcelona, Spain) in 20 minutes. Acetic acid (0.1%) in double distilled water was used as carrier solvent. The flow rate was 0.2 mL/min with injection volumes of 7.5 µl. Sample compounds were identified with mass spectroscopy. The mass spectrometer (esquire HCT, Bruker Daltonics, Bremen, Germany) was operated in negative ion mode with a voltage of + 4.6 keV and a desolation gas flow of 8 l/min at 350°C. The monomers were quantified from their extracted ion chromatograms (EIC) at m/z 159 for 3-OH-C8. The monomers 3-OH-C6, 3-OH-C7:1, 3-OH-C9:1 and for 3-OH-C11:1 were analyzed at m/z 131, m/z 143, m/z 171 and m/z 199.

Purification of 3-hydroxyalkanoic acids

One liter of cell suspension was centrifuged at 5,000 x g and the supernatant adjusted to pH 1 with concentrated HCl. The hydrophobic compounds of this supernatant were then bound to a C18 reversed phase solid phase extraction cartridge (Macherey Nagel, Chromabond HR-P), washed with the five fold cartridge volume of 0.1 M HCl and eluted with acetonitrile. The solvent was evaporated to give 650 mg of a crude mixture containing approximately 60% of 3-hydroxy octanoic acid (determined by HPLC-ESI-MS).

The above mixture was dissolved in 5 mL acetonitrile and subjected to an additional purification step on a RP-18 semi-preparative chromatography column (10 x 250 mm, 7 µm, RP18, Macherey-Nagel Inc., Easton, PA). Separation was achieved using a linear gradient from 0% to 50% acetonitrile (Sharlau, Multisolvent grade, Barcelona, Spain) in double distilled water within 16 minutes. The flow rate was 2 mL/min with injection volumes of 500 µL. The fractions containing 3-hydroxyoctanoic acid were pooled and the solvent evaporated. 280 mg 3-hydroxyoctanoic acid with a purity of more than 95% was obtained (confirmed by ^{1}H-NMR and HPLC-ESI-MS, data not shown).

Polymer isolation and chirality analysis of monomers

PHA was extracted directly from the lyophilized cells (1 mbar, 48 to 144 h). Cells were transferred into pure methylene chloride (60 g CDW in 1 L methylene chloride). After the suspension was stirred overnight, the solution was filtered and concentrated by distillation at 60°C until the solution became viscous. The polymer was then precipitated into ice-cold methanol (final ratio of CH_2Cl_2/MeOH was 1:6 (v/v)). After removal of the solvents by filtration, the PHA was vacuum-dried (30°C, 30 mbar) for at least 1 day and stored at room temperature.

Purified polymers, commercially purchased R,S-3-hydroxyoctanoic acid (Sigma-Aldrich, Seelze, Germany) and the monomer purified in this study were methylated.[256] The absolute configurations of the 3-hydroxyoctanoic acid methyl esters obtained were determined by GC performed with a type Beta-DEX 120 column (fused silica capillary column; length, 30 m; inside diameter, 0.25 mm; film thickness, 0.25 µm; Supelco, Switzerland). The temperature profile started with an isothermal oven temperature of 115°C for 15 min; then the temperature was increased from 115°C to 122°C at a rate of 0.5°C/min, followed by an additional isothermal period of 15 min at 122°C. Compounds were detected by a FID using hydrogen as a carrier gas. R and S enantiomers were identified on the basis of retention times of the methyl esters of commercially available R,S-3-OH-octanoic acid (Sigma-Aldrich, Seelze, Germany) and R-3-hydroxyoctanoic acid obtained from purified PHA of P. putida GPo1.

Reproducibility

In this study we have tested different cultivation conditions, such as C/N of 15 (g/g) and 19 (g/g). For different cultivations, the absolute values regarding monomer production and PHA degradation had variations due to different cell dry weights and intracellular PHA contents; however, the in vivo degradation of PHA was influenced by the environmental conditions in the same manner in all experiments. For the same cultivation the experiments were performed in duplicates. The standard deviation (SD) of the data obtained from GC analysis was ± 5%, from HPLC-MS analysis was ± 10%.

Results and Discussion

1. Selection of a model and a control strain for *in vivo* depolymerization of PHA

Pseudomonas putida strains are one of the most attractive candidates for studying biosynthesis of chiral synthons. Firstly, they are able to accumulate PHA up to 63% (w/w) of cell dry weight (CDW).[257] Secondly, they have a broad carbon substrate spectrum, and thirdly, the PHA depolymerase-deficient mutants of *P. putida* are available.

The wild-type strains *P. putida* GPo1, *P. putida* U and their corresponding depolymerase-deficient mutants *P. putida* GPo500 and *P. putida* U *phaZ*⁻ were grown in batch culture on modified medium E2 containing octanoate as sole carbon source. It was found that both wild-type strains started to degrade PHA and to produce 3-hydroxyalkanoic acid monomers when the cells entered the stationary growth phase, and the carbon source was depleted (data not shown). Similar phenomena were observed by Hayward *et al.*[258] for *Rhizobium*, *Spirillum*, and *Pseudomonas* species. In contrast, no monomer release was detected for GPo500 during the whole time period, indicating that GPo500 was not able to degrade PHA, similar to what was reported previously.[254] Thus, GPo500 can be used as a negative control for *in vivo* depolymerization of PHA. Surprisingly, the depolymerase-deficient mutant *P. putida* U *phaZ*⁻ was able to release 3-hydroxyalkanoic acid monomers, even though at a lower level (approx. 30%) compared to its parent wild type U. This data suggests that mutant U *phaZ*⁻ still contains a PHA depolymerase activity and is not suitable as a control strain. Therefore, *P. putida* GPo1 and GPo500 were used throughout the study.

So far, not much is known about the biochemistry of intracellular PHA depolymerases, especially of mcl-PHA depolymerases. Previously Solaiman et al. have reported that depending on the site of transposon insertion, the *phaZ* gene may or may not be completely inactivated.[259] The *P. putida* U *phaZ*⁻ was generated by inserting transposon into the amino acids between 86-228 of *phaZ*.[38] The putative lipase box was removed. However, we cannot rule out the possibility that the produced fusion protein still contains partial depolymerase activity. Another possibility for the observed monomer production in *phaZ*⁻ mutant is that other depolymerases exist in addition to PhaZ, although mcl-*phaZ* isogenes were not found or reported in *P. putida*. Further studies are necessary to address this issue.

2. Optimization of *in vivo* depolymerization of PHA

To achieve good reproducibility, continuous cultures were used in the following experiments. *P. putida* GPo1 was grown under double-nutrient-limited growth conditions with octanoate as sole carbon source. Under such growth conditions the cells efficiently convert carbon into PHA and biomass,[53] and all carbon source is completely consumed, which will dramatically simplify the downstream processing of monomer extraction and purification. Cells with 1.2 g cell dry weight (CDW)/L and 40% (w/w) PHA relative to CDW were collected. PHA assays revealed that the PHA consisted of 87% 3-hydroxyoctanoic acid (3-OH-C8) (w/w) and 13% 3-hydroxyhexanoic acid (3-OH-C6) (w/w).

2.1 *In vivo* depolymerization of PHA in water

It has been reported that using water as a medium could lead to effective depolymerization of PHB,[70] we thus tested water as medium for mcl-PHA degradation. GPo1 cells described above were resuspended in distilled water to 1.2 g CDW/L, and incubated at 30°C. The cells started immediately to release monomers (mainly 3-OH-C8), and reached a maximum of 46 μg/mL 3-OH-C8 after 6 h (Fig. 2.2).

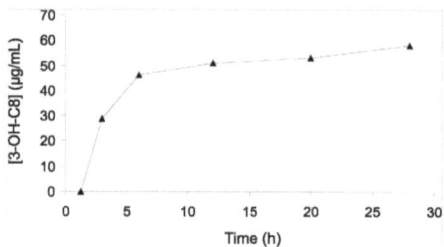

Figure 2.2 In vivo depolymerization of PHA in water. Cells with 1.2 g CDW/L and 40% PHA (w/w) were resuspended in distilled water and incubated at 30°C. The time course of 3-hydroxyoctanoic acid production was followed for about 30 h. The experiment was performed in duplicate. The standard deviation (SD) was ±10%.

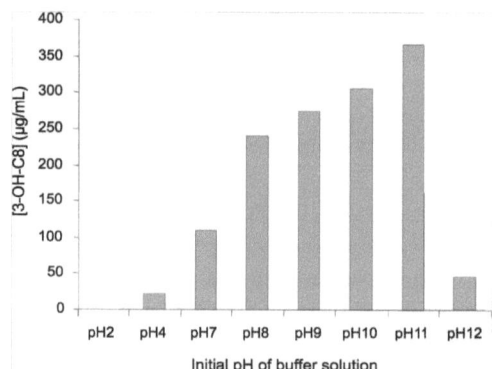

Figure 2.3 Influence of initial pH on in vivo depolymerization of PHA. Cells were resuspended in 50 mM phosphate buffer with an initial pH of 2, 4, 7, 8, 9, 10, 11, or 12 and were incubated at 30°C. Samples were taken after 6 h of incubation for monomer analysis. The experiment was performed in duplicate. The SD was ±10%.

Further incubation did not lead to much higher monomer amount. The pH of the reaction mixture decreased from initial 6.7 to around 6.3 after 6 h. In the control strain GPo500 no monomer release was detected and no significant pH change was observed. We reasoned that the low monomer production yield (maximal theoretical yield of 3-OH-C8 was 418 μg/mL based on 1.2 g CWD/L and 40% PHA containing 87% 3-OH-C8) could be caused by the low pH, since it was reported that the optimal pH range for intracellular PHA depolymerases is at 8-10 (see review by Jendrossek and Handrick[15]). Thus, we further investigated the effects of different initial pH on *in vivo* depolymerization.

2.2 Influence of pH on *in vivo* depolymerization of PHA

Cells described above were collected by centrifugation and resuspended to 1.2 gL^{-1} in 50 mM phosphate buffer with different initial pH: pH 2, pH 4, pH 7, pH 8, pH 9, pH 10, pH 11 and pH 12, and were incubated at 30°C. The amount of the monomer released was measured after 6 h. Figure 2.3 shows that the initial pH had a significant effect on *in vivo* monomer release: the highest amount of monomers was obtained with the initial pH 11, where 360 µg/mL 3-OH-C8 was obtained. Increase of the initial pH to 12 led to a small amount of monomers (below 50 µg/mL). At an initial pH of 10 the produced 3-OH-C8 concentration was reduced to 300 µg/mL. Further decrease of the initial pH resulted in further reduction of 3-OH-C8 production (Fig. 3). At an initial pH of 2 no detectable monomers could be measured. In addition to 3-OH-C8, 3-OH-C6 was also detected. Since purified 3-hydroxyhexanoic acid was not available as a standard, the produced 3-hydroxyhexanoic acid could not be quantified.

It is not likely that the monomers were produced by chemical hydrolysis reaction. Firstly, by using the initial pH 12 the amount of released 3-OH-C8 decreased dramatically (Fig. 2.3). If the reaction were alkaline hydrolysis, there would be no reason why it took place only at pH 8-11, not at pH 12. Secondly, when the depolymerase-deficient mutant GPo500 was analyzed for monomer release at pH 8-11 in batch culture, there was no/small amount of monomers were found (data not shown). Thirdly, to inactivate depolymerase cells were heated at 70°C for 3 min before collection, and then collected and resuspended in 50 mM phosphate buffer with initial pH 11 as described above. The heated cells did not release any trace of monomers, suggesting that inactivation of depolymerase abolished the ability of PHA degradation. Thus, the obtained monomers were produced due to *in vivo* depolymerization, not to chemical hydrolysis.

2.3 Time profile of PHA depolymerization

Since the above observed monomer production was only an endpoint measurement, the time profiles of *in vivo* depolymerization at selected pH (pH 7, pH 10, pH 11 and pH 12) were further examined (Fig. 2.4). At all tested pH the production of 3-OH-C8 was close to the maximum already after 6 h of incubation (Fig. 2.4A). Further incubation up to 12 hours led to only slightly higher concentrations. The optimal initial pH for *in vivo* depolymerization was found once more at pH 11 (Fig. 2.4A). Besides 3-OH-C8, the release of 3-OH-C6 monomer was observed (Fig. 2.4B). Based on the peak areas from HPLC, we found that the production of 3-OH-C6 followed a similar pattern to that of 3-OH-C8 (Fig. 2.4).

We also found that the pH did not change much during the incubation when the cells were resuspended in phosphate buffer with initial pH of 7 or 12, whereas with initial pH of 10 and 11 the pH of the culture mixture changed significantly (Fig. 2.4C): pH 11 was decreased to 8.8 and pH 10 to 7.8 after 6 hours. It has been reported that PHA can serve as carbon and energy reservoirs, which enhance survival and stress tolerance in bacteria.[159] Our data suggest that depolymerization may also be a mechanism to adjust the extracellular pH of the cells.

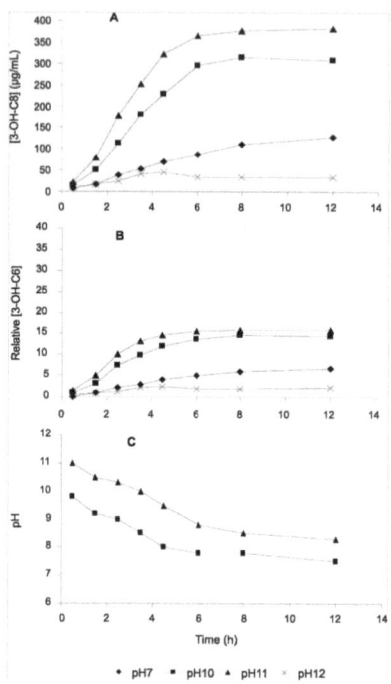

Figure 2.4 Time profile of saturated 3-hydroxyalkanoic acid production at different pH values. Cells with 40% PHA (w/w) were collected and resuspended in 50 mM phosphate buffer with an initial pH of 7, 10, 11, or 12 and were incubated at 30 °C. Samples were taken at different time intervals for analysis of monomer production. The experiments were performed in duplicate. (A) Production of 3-OHC8; (B) production of 3-OH-C6 (based on the HPLC peak areas in relation to 3-OH-C8); (C) pH change during incubation. The SD of the data in panels A and B was ±10% and that in panel C was ±4%.

We also analyzed the cell viability at different pH (from pH 7 to pH 12) under an optical microscope. We observed that after 6 hours there was no significant difference between the cells exposed to pH 7-11 regarding cell morphology and viability. Above pH 11 the cell lysis was observed (data not shown).

Unlike what has been reported previously that intracellular PHB depolymerase from *Alcaligenes latus* was most active under acidic conditions (pH 3-4),[70] the intracellular mcl-PHA depolymerase from *P. putida* in our study was most active under basic conditions. Our observation is in agreement with what was found by Saito and co-workers that intracellular PHA depolymerase of *C. necator* has highest activity at pH 8-10.[220] Indeed *in vitro* studies of the intracellular depolymerases from different organisms have shown that the optimal pH of most PHA depolymerases lies in the alkaline range between 8 and 10.[15] So far the native intracellular mcl-PHA depolymerase from *P. putida* has neither been purified nor characterized. Our data indicate that the optimal pH for this enzyme is also in the basic range.

3. Efficiency of *in vivo* depolymerization of saturated mcl-PHA

To study in detail the efficiency of PHA degradation and monomer production, cells with 1.2 g CDW/L and 40% PHA (%CDW) (87% 3-OH-C8 (w/w) and 13% 3-OH-C6 (w/w)) obtained from continuous culture were collected and resuspended to 1.2 gL^{-1} in phosphate buffer (pH 11). Samples were taken regularly for monomer and PHA measurement. Figure 2.5 shows that PHA immediately started to be degraded after incubation, and correspondingly monomers were produced rapidly. This suggests that the *in vivo* depolymerization reaction can occur very quickly when the bacteria are exposed to appropriate conditions. After 3 hours, half of the PHA was degraded, and 190 µg/mL 3-OH-C8 was produced. After 6 hours most of the PHA was degraded to monomers (356 µg/mL 3-OH-C8) and GC analysis revealed that only 5% PHA (w/w) remained in the cells (Fig. 2.5).

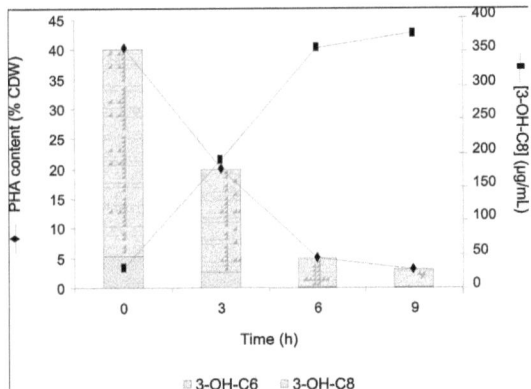

Figure 2.5 Efficiency of 3-hydroxyalkanoic acid production from PHA. Cells with 1.2 g CDW/L and 40% PHA (w/w) were collected and resuspended in 50 mM phosphate buffer with pH 11 and were incubated at 30 °C. Time profiles of PHA degradation and monomer production were followed over 9 h. Bars, PHA composition; dotted bars, 3-OH-C6; striped bars, 3-OH-C8. The experiments were performed in duplicate. The data for PHA content had a SD of ±5%, whereas that for monomer content has a SD of ±10%.

The monomer composition of 3-OH-C8 and 3-OH-C6 of the intracellular PHA was changed from 87:13 (w/w) to 98:2 (w/w) after 6 hours, while the relative ratio of the released 3-OH-C8 and 3-OH-C6 via depolymerization was 96:4 (w/w). It is possible that part of the released 3-OH-C6 was re-utilized by the cells. However, we cannot rule out the possibility that this difference was caused by different substrate affinities of PHA depolymerase. After 9 hours only little PHA (3% relative to CDW) was detected in the cells, and 378 µg/mL 3-OH-C8 was released. After 24 hours no detectable PHA was left in the cells. Thus, PHA degradation efficiency and monomer production yielded over 90% (w/w). Compared to what was reported previously by Lee and co-workers that the production yield of mcl-3-hydroxyalkanoic acids could be reached at maximal 9.7% after 4 days,[70] the yield obtained in this study is much higher.

Chapter 2

4. Production of unsaturated monomers

To assess whether the *in vivo* depolymerization developed in this study can also be applied to prepare other types of *R*-hydroxyalkanoic acids, PHA with unsaturated monomers were tested. *P. putida* GPo1 was cultivated in continuous culture supplemented with 10-undecenoic acid. Cells with 1.5 g CDW/L and 37% PHA were obtained and collected. The PHA was composed of 10% 3-OH-6-heptenoic acid (3-OH-C7:1), 62% 3-OH-8-nonenoic acid (3-OH-C9:1) and 28% 3-OH-10-undecenoic acid (3-OH-C11:1) (w/w). The cells were resuspended to 1.5 g CDW/L in 50 mM phosphate buffer with different pH, and incubated at 30°C. Since the pure compounds 3-OH-C7:1, 3-OH-C9:1 and 3-OH-C11:1 were not commercially available, we could only calculate the relative amount of the obtained monomers based on the HPLC peak areas. Figure 2.6A shows that the influence of pH on the release of unsaturated monomers exhibited a similar pattern to that on saturated monomers (Fig. 2.3).

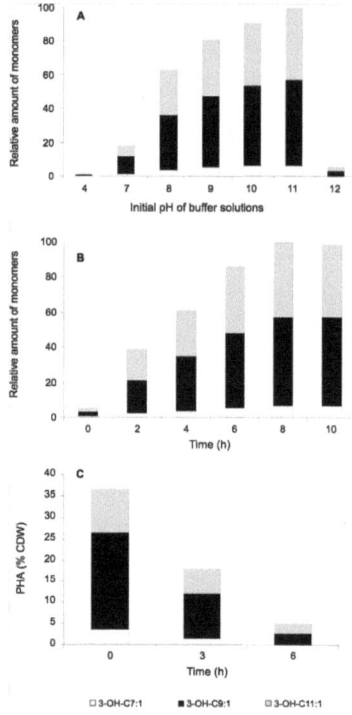

Figure 2.6 Production of unsaturated monomers. Cells with 1.5 g CDW/L and 37% PHA (w/w) were collected and resuspended in 50 mM phosphate buffer with different pH. (A) Influence of pH on monomer production. 100% corresponds to the highest amount of total monomers obtained. (B) Time profile of unsaturated monomer production in buffer with pH 11. 100% corresponds to the highest amount of total monomers obtained. (C) Unsaturated PHA degradation in buffer with pH 11. The SD of the data in panels A and B was ±10% and that in panel C was ±5%.

The buffer with initial pH 11 was found to provide the best condition also for the unsaturated monomer production. The relative weight ratio of the obtained 3-OH-C7:1, 3-OH-C9:1 and 3-OH-C11:1 based on the peak areas was 6:52:42 after 6 hours. The rate of *in vivo* depolymerization of unsaturated PHA also followed a similar manner to that of saturated PHA (Fig. 2.4 and 2.6B): namely the monomer release took place immediately after incubation and reached the maximum after 8 hours (Fig. 2.6B). This was slightly longer than what was needed for the monomer release of saturated PHA (Fig. 2.4).

We also studied the efficiency of olefinic PHA degradation and monomer production (Fig. 2.6C). Although we could not quantify the absolute amount of obtained monomers by HPLC, the GC analysis revealed that PHA was degraded from initially 37% (w/w) to about 5% (w/w) after 6 h. After 24 hours no detectable PHA was left in the cells. Thus, we conclude that the developed conditions were also suitable for unsaturated monomer production.

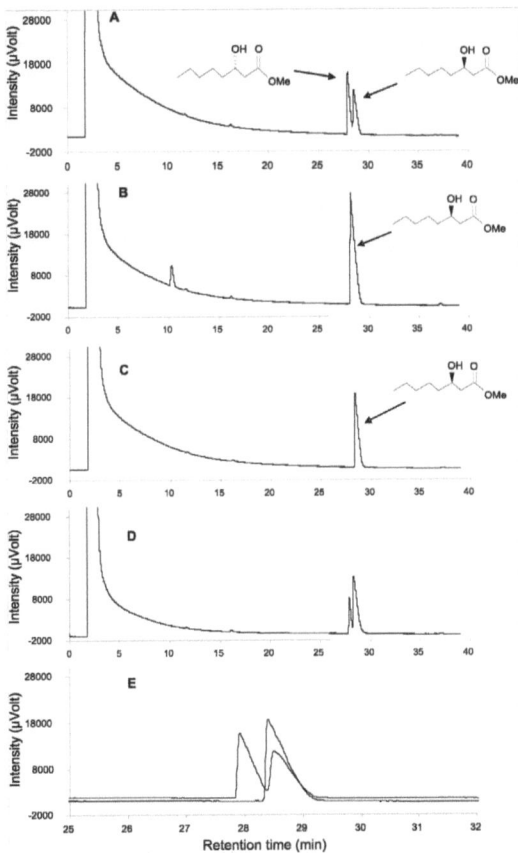

Figure 2.7 Chiral gas chromatography analysis of 3-hydroxyoctanoic acids produced by *in vivo* depolymerization. (A) Racemic standard mixture of R,S-3-hydroxyoctanoic acid methyl esters; (B) 3-hydroxyoctanoic acid methyl ester prepared from purified PHA isolated from *P. putida* GPo1; (C) methyl ester of 3-hydroxyoctanoic acid produced and purified in this study; (D) mixture [1:1 (v/v)] of the racemic standard (panel A) and 3-hydroxyoctanoic acid methyl ester prepared from in vivo depolymerization (panel C); (E) comparison of R-3-hydroxyoctanoic acid methyl ester with the racemic mixture by superimposition of the chromatograms.

5. Chirality of monomers produced via *in vivo* depolymerization

Since the obtained monomers were a mixture of co-monomers in phosphate buffer, we further separated and purified these monomers on reversed phase semi-preparative liquid chromatography as described in Experimental Section. The purified 3-OH-C8 was analyzed for chirality. It has been demonstrated that PHA monomers produced by *P. putida* GPo1 have only the *R* configuration.[256] PHA containing 3-OH-C8 and 3-OH-C6 purified from GPo1 were methanolyzed to methyl esters according to the PHA assay reported previously.[256] The same treatment was applied to the commercially available 3-OH-C8 (Sigma) and the monomers purified in this study. The esters were analyzed by chiral gas chromatography. Two controls were used: methyl ester of the racemic 3-OH-C8 (Sigma) (Fig. 2.7A) and methyl ester of *R*-3-OH-C8 obtained by hydrolyzing PHA (Fig. 2.7B). For the racemic mixture two peaks with retention times of 27.93 min (left) and 28.51 min (right) were detected (Fig. 2.7A), where the peak at 28.51 min corresponded to the *R*-enantiomer obtained from the purified PHA (Fig. 2.7B). The monomer produced via depolymerization in this study yielded a single peak with retention time of 28.42 min (Fig. 2.7C), confirming the presence of only one enantiomer. Further mixing racemic mixture (Fig. 2.7A) and purified monomers (Fig. 2.7C) increased the right peak at 28.51 min (Fig. 2.7D), clearly demonstrating that the monomers purified in this study has exclusively *R*-configuration. Superimposition of chromatograms of Figure 2.7A and Figure 2.7C suggests that the isolated and purified monomer has *R*-configuration with very high enantiomeric excess (Fig. 2.7E).

Conclusions

In this study, we have demonstrated that various enantiomerically pure *R*-3-hydroxyalkanoic acids can be produced via *in vivo* depolymerization of PHA. The cells were first grown under double (carbon and nitrogen) nutrient limitation conditions for a high PHA accumulation and afterwards resuspended in phosphate buffer adjusted to basic environments. The production yield of 3-hydroxyalkanoic acids reached over 90% in 9 h under these conditions, which is significantly higher than what has been reported previously (9.7% yield in 4 days).[70] More importantly, phosphate buffer as a production medium for the monomers leads to effective cost reduction, easy downstream processing, and an environmentally friendly approach. The method developed here might be a first step towards large scale production of different chiral *R*-hydroxyalkanoic acids.

Since mcl-PHA synthesis in *P. putida* GPo1 has a broad substrate range, we expect that many of as-yet-untested hydroxyalkanoic acid intermediates can be incorporated into mcl-PHA, which can then be *in vivo* depolymerized to *R*-hydroxyalkanoic acids by using the strategy shown in this study. Thus, this work opens a new route for the production of various enantiomerically pure chemicals.

Acknowledgement

We thank Elisabeth Michel, Axel Ritter and Jie Zhang for helping with the GC analysis and chirality analysis. We greatly acknowledge the financial support by EMPA.

Chapter 3

Efficient production of (*R*)-3-hydroxycarboxylic acids by biotechnological conversion of polyhydroxyalkanoates and their purification

Katinka Ruth, Andreas Grubelnik, René Hartmann, Thomas Egli, Manfred Zinn, and Qun Ren. 2007.
Biomacromolecules, 8 (1), 279-286.

Abstract

An efficient method to prepare enantiomerically pure *(R)*-3-hydroxycarboxylic acids from bacterial polyhydroxyalkanoates (PHA) accumulated by *Pseudomonas putida* GPo1 is reported in this study. *(R)*-3-hydroxycarboxylic acids from whole cells were obtained when conditions were provided to promote *in vivo* depolymerization of intracellular PHA. The monomers were secreted into extracellular environment. They were separated and purified by acidic precipitation, preparative reversed-phase column chromatography, and subsequent solvent extraction. Eight *(R)*-3-hydroxycarboxylic acids were isolated: *(R)*-3-hydroxyoctanoic acid, *(R)*-3-hydroxyhexanoic acid, *(R)*-3-hydroxy-10-undecenoic acid, *(R)*-3-hydroxy-8-nonenoic acid, *(R)*-3-hydroxy-6-heptenoic acid, *(R)*-3-hydroxyundecanoic acid, *(R)*-3-hydroxynonanoic acid, and *(R)*-3-hydroxyheptanoic acid. The overall yield based on released monomers was around 78 weight% for *(R)*-3-hydroxyoctanoic acid. All obtained monomers had a purity of over 95 weight%. The physical properties of the purified monomers and their antimicrobial activities were also investigated.

Introduction

Polyhydroxyalkanoates (PHA) are biodegradable and biocompatible polyesters which are produced by a wide variety of microorganisms.[21, 174] PHA are stored in intracellular granules and used by bacteria as carbon and energy source.[174] Many bacteria accumulate PHA when they are cultured in excess of carbon source and when growth is limited by the lack of one or more essential nutrients.[21, 166] Under carbon starvation the accumulated PHA can be degraded to monomers which can be utilized for better cell survival.[15] To date, more than 140 types of 3-, 4-, 5-, or 6-hydroxycarboxylic acids have been found to be incorporated into the polymer.[13, 78] All monomers analyzed so far have absolute *(R)*-configuration.[92, 256]

(R)-3-hydroxycarboxylic acids (R3HA) have been reported to be valuable synthons and can be used as starting materials for synthesis of antibiotics, vitamins, flavors, and pheromones.[80, 81, 260-262] It was also reported that some of *(R)*-3-hydroxycarboxylic acids exhibit important biological activities, such as antimicrobial and/or antiviral potential.[224, 263] For example, *(R)*-3-hydroxy-*n*-phenylalkanoic acids can be used to effectively attack *Listeria monocytogenes*, which is a ubiquitous microorganism, able to multiply at refrigeration temperatures, and is resistant to both high temperatures and low pH values.[224] This food pathogen has become an important issue in many countries during the past several decades.

Different methods were described to produce R3HA: They can be chemically synthesized by enantioselective reduction of the corresponding 3-keto acids.[84] For certain products, this process requires the synthesis of precursor molecules, which complicates the synthetic procedure and may reduce the product yield.[86, 87] Other synthetic approaches include stereoselective functionalization through Sharpless' asymmetric epoxidation and hydroxylation or through Brown's asymmetric allyboration.[82] These approaches require chiral, often expensive metal-complex catalysts which might contaminate the final products. For some conversions vigorous reaction conditions have to be applied such as high pressure, flammable reaction media, or cryogenic conditions.[82, 85] Furthermore, products often have lower enantiomeric excess than those obtained by biochemical processes.[70] Other approaches include microorganisms such as recombinant *Escherichia coli*[89] or *Saccharomyces cerevisae* as biocatalysts to introduce the chiral center.[90] Product inhibition of enzyme activity and moderate yields of product isolation can be limitations of these procedures.[70, 81]

Alternative to chemical synthesis, bacterial PHA could be used as an important source of *R*3HA.[80] In order to obtain *R*3HA from PHA, *in vivo* depolymerization has been investigated and efficiently accomplished for poly(3-hydroxybutyrate).[70] *(R)*-3-hydroxybutyrate has been produced with a yield of 96%.[70] *R*3HA with chain lengths of 6-11 carbon atoms can also be produced with a high yield (90%) by *in vivo* depolymerization of PHA from *Pseudomonas putida* GPo1.[264] However, bacterially produced PHA is often a co-polymer containing *R*3HA with chain length differing in one or more ethylene units.[78] The corresponding monomers derived from one type of PHA have very similar chemical properties and, therefore, they are difficult to separate.

One approach to purify a mixture of different *R*3HA generated from PHA has been reported by de Roo and co-workers:[92] *R*3HA methyl esters were produced by acid-catalyzed hydrolysis of the extracted PHA. The distinct methyl esters were separated by fractional distillation and the free acids were obtained by subsequent saponification. Even though high yields were obtained (~80%), this procedure is rather complicated and tedious. It takes 7 days for distillation. Furthermore, considerable amounts of organic solvents are needed for the extraction of PHA.

A satisfying process to provide *R*3HA on a preparative scale has not yet been achieved. For future applications of *R*3HA as synthons or antimicrobial agents, it is not only important being able to produce *R*3HA, but also to isolate them to a high degree of purity. In this study we developed an easy process to produce various chemically pure *R*3HA in mg to g quantities as shown in Figure 3.1. *P. putida* GPo1 in continuous growth culture under dual-nutrient-limited conditions (C- and N-limitation) was used to accumulate PHA. Afterwards, the bacteria were exposed to an environment where enzymatic depolymerization of intracellular PHA took place. PHA monomers are secreted, thus extraction of PHA with organic solvents is not necessary which renders this step environmentally friendly. The monomeric units of the PHA were collected and separated by reversed phase preparative column chromatography followed by solvent extraction. The overall yield based on released monomers was around 78 weight% for *(R)*-3-hydroxyoctanoic acid. The physical properties of the purified monomers and their antimicrobial activities were also investigated.

Materials and methods

Continuous cultivation

In order to produce PHA, the wild-type strain *P. putida* GPo1 (ATCC 29347) was cultivated in chemostat cultures at a dilution rate (D) of 0.1 h^{-1} under dual-nutrient-limited growth conditions.[31, 63] Continuous cultivation was performed as previously reported:[265] 40 L of continuous culture medium supplied with mineral trace elements (CCMT) was filter-sterilized (0.22 μm filter) into γ-sterilized 50-L medium bags (Flexboy, Stedim S. A., Aubagne Cedex, France). The carbon source (octanoic, undecanoic or 10-undecenoic acid (Fluka, Buchs, Switzerland)) was pumped directly into the culture vessel by using a dosimat (Metrohm, Herisau, Switzerland). A 3.7 L laboratory bioreactor (KLF 2000, Bioengineering, Wald, Switzerland) with a working volume of 2.8 L was used. Cells grew in CCMT supplied with octanoic acid or 10-undecenoic acid at a carbon to nitrogen ratio (C/N) of 15.0 g·g^{-1}, or with undecanoic acid at a carbon to nitrogen ratio (C/N) of 10.8 g·g^{-1}. The cultures were kept at 30 ± 0.1°C, and the pH was maintained at 7.0 ± 0.05 by automated control. The dissolved oxygen tension was monitored continuously with an oxygen probe (Mettler Toledo, Greifensee, Switzerland) and care was taken that it remained above 35% air saturation. The culture volume was kept constant with a balance that controlled the harvest pump. The harvested cell broth was collected in a 10 L tank kept on ice.

Ammonium was determined by a photometric ammonium assay (Spectroquant, Merck, Darmstadt, Germany). The carbon content of the medium was controlled with a total organic carbon analyzer (TOC-5050A, Shimadzu, Reinach, Switzerland). The cell dry weight (CDW) was recorded as described previously.[31]

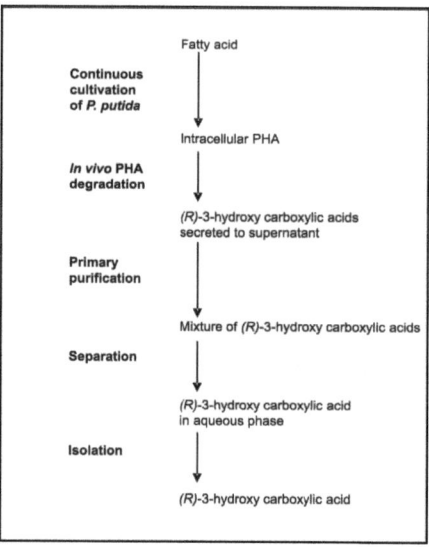

Figure 3.1 Schematic representation of the process to isolate (R)-3-hydroxycarboxylic acids (R3HAs).

Monomer production

Cells collected during steady-state conditions were resuspended in 50 mM phosphate buffer at pH 10 to a concentration of 3-10 g·L^{-1} depending on the CDW and PHA content in order to achieve a final PHA concentration of around 1.5 g·L^{-1}. The suspension was incubated at 30°C for 8-10 h. Subsequently, the supernatant was obtained by centrifugation (4,500 x g, 4°C, 20 min; Multifuge 3 S-R, Osterode, Germany) or filtration (filters LS14 ½ 270 mm, Schleicher & Schuell, Feldbach, Switzerland). Finally, the supernatant containing the PHA monomers was acidified to pH 1 with concentrated HCl. The monomers remained in the supernatant and were subjected to the separation process. It was possible to store the monomer-containing solution at 4°C for several months without degradation of monomers (controlled by HPLC-MS).

Monomer separation

Separation of various R3HA was first attempted by solvent extraction methods. However, the amphiphilic character of R3HA led to formation of an emulsion. Complete phase separation did not occur and R3HA accumulated in the inter-phase. Several anion exchange resins for column chromatography were tested as well, but the separation of R3HA with these resins was not satisfactory. The reason for this presumably is that ionic properties of all R3HA are too similar.

Monomer separation was accomplished with a glass column (length 50 cm; diameter 1.5 cm) packed with 40 g silica gel 100 C18-reversed phase (particle size 0.040-0.063 mm; max. surface coverage 17-18% C or ± 4 µmol·m^{-2}; Fluka). During operation, the column was cooled to <10°C and set under N$_2$-pressure (~80 kPa). After washing with 0.1 M HCl (> 5x bed volume), the mixture of monomers was fractionated using a stepwise elution with 0.1 M HCl/acetonitrile mixtures as mobile phase. Firstly, 100-150 mL of 0.1 M HCl/acetonitrile with a composition of 85/15 (vol%) were applied, secondly 300-400 mL of 0.1 M HCl/acetonitrile with a composition of 50/50 (vol%), and finally 100 mL of pure acetonitrile. The collected fractions were analyzed for the presence of monomers by high performance liquid chromatography (HPLC).

Solvent extraction

The fractions containing only one type of R3HA were pooled and the acetonitrile was removed by rotary evaporation (150 mbar; 60°C). Subsequently, the solutions were further acidified by addition of 1 M HCl (acid/sample = ~1/1 (vol%)), saturated with KCl, and extracted three times with the equal volume of tert-butyl methyl ether (aqueous phase:organic phase = 1/1 (vol%)). The combined organic phases were dried over Na$_2$SO$_4$, filtered and the solvent evaporated (450 mbar; 40°C). The purified monomers were stored under vacuum at room temperature.

Antimicrobial assay

The antimicrobial activity of R3HA was tested by measuring the effect caused by several concentrations of the purified R3HA (ranging from 100 µM to 10 mM) on the growth of different species of bacteria (*Listeria innocua, Listeria monocytogenes, Listeria ivanovii, Escherichia coli* MG1655, *Salmonella enterica, Staphylococcus aureus* RN4220). (R/S)-3-hydroxyoctanoic acid (Larodan, Malmö, Sweden) (ranging from 100 µM to 10 mM) and octanoic acid (ranging from 100 µM to 100 mM) were also investigated for comparison. Other racemic mixtures such as (R/S)-3-hydroxy-6-heptenoic acid, (R/S)-3-hydroxy-8-nonenoic acid, and (R/S)-3-hydroxy-10-decenoic acid were not commercially available, thus the experiments using these compounds could not be carried out. The bacteria were cultured in TSB media (17 gL^{-1} casein peptone, 3 gL^{-1} soya peptone, 5 gL^{-1} NaCl, 2.5 gL^{-1} K$_2$HPO$_4$, 2.5 gL^{-1} glucose) in the presence or in absence of the molecule to be tested. The pH was adjusted to pH 7.0 with sodium hydroxide. Conditions were 37°C and shaking with 150r/min. Precultures were grown in TSB to exponential phase. For inoculation, they were diluted 1 to 30 (vol/vol) into the test solution to start the assay. Their growth was determined by measuring the optical density of the cultures at 550 nm over 24 hours, using a microtiter plate procedure.[224, 266] Minimal inhibitory concentration (MIC) was determined as the lowest concentration where no bacterial growth was observed.

Analytical Methods

Gas chromatography (GC). Quantitative analysis of R3HA was carried out by applying the following method adapted from a procedure for fatty acids:[267] For monomer measurement, 8 mL of monomer-containing solution or 5-15 mg isolated monomers were filled into a 10 mL Pyrex tube. In the former case the liquid phase was evaporated by a nitrogen stream at 175 mL·min^{-1}. 1 mL of 2-ethyl-2-hydroxybutyric acid (Fluka), dissolved in dichloromethane, was added as internal standard. Its concentration was set to be similar to the estimated monomer content (1-10 mg/mL). Subsequently, 1 mL of BF$_3$ (boron trifluoride; ~10% in methanol; Fluka) was added and the sample was tightly capped and heated to 80°C for 20 hours. Afterwards, 2 mL of dichloromethane and 2 mL of saturated NaCl solution were added, and after intense shaking the upper phase was discarded and the lower one dried over Na$_2$SO$_4$ and subsequently analyzed on a GC (Fisons Instruments, Rodano, Italy) equipped with a polar fused silica capillary column (Supelcowax-10: length 30 m; inside diameter 0.31 mm; film thickness 0.5 µm; Supelco, Buchs, Switzerland). The injection temperature was 250°C; the injection volume was 1 µL with a split ratio of 1:10. Helium was used as the carrier gas (3 mL·min^{-1}) and detection was performed with a flame ionization detector (FID) at 285°C. The temperature was increased from 80 to 280°C at a rate of 10°C·min^{-1} to record a GC-spectrum. The content was calculated with regard to the signal intensity of the internal standard. The calibration was accomplished with racemic 3-hydroxycarboxylic acids of varying chain lengths (Biotrend, Köln, Germany). The detected response factors were used to calculate absolute concentrations of samples.

For intracellular PHA measurement, the cell suspension was centrifuged (10,000 x g; 4°C; 15 min) and the cell pellet was lyophilized for 48 hours. Around 50 mg of dry biomass were filled into a 10 mL Pyrex tube and treated in the same manner as for the monomers.

Chiral GC analysis. Purchased (R,S)-3-hydroxycarboxylic acids (Biotrend, Köln, Germany) and R3HA purified in this study were methylated.[256] Separation of the enantiomers was performed on a GC equipped with a Beta-DEX 120 column (fused silica capillary column; length 30 m; inside diameter 0.25 mm; film thickness 0.25 µm; Supelco, Buchs, Switzerland). Absolute configurations of the methyl esters were determined by comparison with reference materials. The injection volume was 1 µL with a split ratio of 1:10. The temperature was increased from 100 to 130°C at a rate of 1°C min^{-1} for optimal peak separation. The carrier gas was helium and the compounds were detected by FID.

HPLC. Samples were diluted to 0.1-10 ppm in acetonitrile/water/acetic acid (50.0/49.9/0.1 (vol%)). Separation of compounds was performed on a reversed phase C18 column (Gemini 5 µm C18 110Å,

250 x 2.00 mm, Phenomenex, Macclesfield, U.K.) applying a linear gradient of 100% diluted acetic acid (0.1 vol% in water) to 100% acetonitrile as the mobile phase. The flow rate was 0.2 mL·min^{-1}. The gradient was completed within 11 minutes. After each run the column was equilibrated at starting conditions for 8 minutes. The injection volume was 7.5 µL. Peaks were detected in negative mode by electrospray ionization mass spectrometry (ESI-MS) (Esquire HCT, Bruker Daltonics, Bremen, Germany).

DSC. Differential scanning calorimetry (DSC) was performed to determine the melting points (Tm) and enthalpies of the purified R3HA. Samples of 8-14 mg PHA were weighed into aluminum pans and analyzed with a DSC 30 (Mettler Toledo, Greifensee, Switzerland). The samples were cooled to -80°C within 10 min. After equilibration of temperature, they were heated to 100°C at a heating rate of 10°C min^{-1}. Melting points were evaluated based on maximum peak heights. Data were determined with a standard deviation of ±13%.

^{1}H NMR. Proton nuclear magnetic resonance experiments in CDCl$_3$ (deuterated chloroform) solution were performed on a Bruker AV-400 spectrometer. Chemical shifts are given in ppm relative to the remaining signals of chloroform as internal reference (^{1}H NMR: 7.26 ppm).

Abbreviations

Three different co-polymers were produced by continuous cultivation of *P. putida* GPo1. They were used to isolate 8 different R3HA. For ease of readability, special abbreviations for the obtained polymers and monomers were used, as listed in Table 3.1.

Results and Discussion

In this study, *P. putida* GPo1 was selected for the biosynthesis of PHA due to its broad carbon substrate spectrum and its ability to accumulate PHA up to 63% of cell dry weight (CDW).[257] *P. putida* GPo1 was grown in continuous cultivation under carbon and nitrogen (C,N) limitation as described in materials and methods. These well-defined dual-nutrient-limited growth conditions ensured optimal and reproducible production of homogenous PHA polymers.[64]

Production of intracellular PHA

Three types of intracellular PHA were synthesized in this study based on the cultivation parameters listed in Table 3.2. When octanoic acid was used as the sole carbon source, 1.47 g·L^{-1} CDW and 45.5 weight (wt) % PHA were obtained. The monomer composition C8-0/C6-0 was found to be 92/8 (wt%). A co-polymer of PHUA was produced with undecanoic acid as carbon substrate. The CDW and PHA content were 6.70 g·L^{-1} and 16.6 wt%, respectively. The monomer ratio of C11-0/C9-0/C7-0 was 9/51/40 (wt%). The third type of polymer, PHUE, was synthesized when 10-undecenoic acid was applied as single carbon source. The CDW of 1.36 g·L^{-1} and the PHA content of 13.5 wt% were detected. The monomer composition of C11-1/C9-1/C7-1 was found to be 14/69/17 (wt%). In order to investigate a potential effect on monomer release patterns, different conditions were applied for PHA accumulation. However, the state of the continuous cultures during PHA accumulation in this context did not seem to play a significant role for subsequent monomer release.

Table 3.1 Abbreviations used for polymers and monomers investigated in this study.

Substance		Abbreviation
Polymers		
poly((R)-3-hydroxyoctanoate-co-3-hydroxyhexanoate)		PHO
poly((R)-3-hydroxyundecanoate-co-3-hydroxynonanoate-co-3-hydroxyheptanoate)		PHUA
poly((R)-3-hydroxy-10-undecenoate-co-3-hydroxy-8-nonenoate-co-3-hydroxy-6-heptenoate)		PHUE
Monomers		
(R)-3-hydroxyhexanoic acid		C6-0
(R)-3-hydroxyheptanoic acid		C7-0
(R)-3-hydroxy-6-heptenoic acid		C7-1
(R)-3-hydroxyoctanoic acid		C8-0
(R)-3-hydroxynonanoic acid		C9-0
(R)-3-hydroxy-8-nonenoic acid		C9-1
(R)-3-hydroxyundecanoic acid		C11-0
(R)-3-hydroxy-10-undecenoic acid		C11-1

The monomer compositions measured here by GC (e.g., C8-0/C6-0 = 92/8 wt%) were slightly different from those previously reported (e.g., C8-0/C6-0 = 87/13 wt%).[31, 264] This might be caused by different analytical methods which mainly vary in the sample preparation for GC analysis. We compared the previously reported derivatization, namely, the formation of propylesters catalyzed by HCl, with the method used in this study, namely, the formation of methyl esters catalyzed by BF_3. With the former method the monomer composition of C8-0/C6-0 was detected to be 87/13, which was the same as what was reported.[31, 264] In this study the latter method showed better reproducibility and less side-products, hence it was used further throughout subsequent experiments.

Production of (R)-3-hydroxycarboxylic acids

For *in vivo* depolymerization, bacterial cells containing various types of PHA were collected and suspended in 50 mM phosphate buffer at pH 10. In the cell suspension the final PHA concentrations were always around 1.5 g·L^{-1}. The cells were then incubated for 8-10 hours at 30°C. In all cases, PHA degradation started immediately and monomers were secreted into the supernatant as described previously.[264] The efficiency of the intracellular PHA degradation reached over 70 wt% after 8 h when testing the cell pellet for remaining PHA content by GC. At the same time, the supernatant was tested for monomer content and the corresponding yields also achieved over 70 wt%. Different CDW and PHA contents during preceding PHA accumulation by continuous cultivation had no significant influence on either PHA degradation or monomer release under the tested conditions.

Table 3.2 PHA synthesis with different carbon sources in *P. putida* GPo1. Data for CDW and PHA content were derived from at least 4 independent measurements.

Cultivation parameters	PHO	PHUA	PHUE
Carbon substrate	octanoic acid	undecanoic acid	10-undecenoic acid
C/N ratio [g·g^{-1}]	15.0	10.8	15.0
Feed nitrogen [mg (NH$_4^+$)·L^{-1}]	150	800	150
Carbon supply [mg (C)·L^{-1}]	1750	6720	1750
Dilution rate [h^{-1}]	0.1	0.1	0.1
pH	7.00 ± 0.02	7.00 ±0.01	7.00 ± 0.02
CDW [g L^{-1}]	1.47 ± 0.23	6.70 ± 0.51	1.36 ± 0.07
PHA content [wt% of CDW]	45.5 ± 2.2	16.6 ± 1.2	13.5 ± 0.6
Monomer composition [wt%]	C8-0/C6-0	C11-0/C9-0/C7-0	C7-1/C9-1/C11-1
	= 92/8	= 9/51/40	= 14/69/17

Monomer yield obtained in this study was lower (~70 wt%) than what we obtained before (~90 wt%).[264] One explanation for this observation is that the concentration of PHA in the phosphate buffer was higher (≥ 2 times) than what was tested in previous studies.[264] Higher initial PHA concentrations lead to theoretically higher monomer concentrations. Previous studies showed that the release of monomers strongly depended on the extracellular pH and was optimal under alkaline conditions.[264] Released monomers constantly lowered the extracellular pH, which can lead to a unfavorable condition for monomer release.[264] Thus, higher PHA concentration could result in lower monomer yield when the extracellular pH was not controlled at the optimum. A system to regulate pH during monomer release might be appropriate to avoid this influence. However, we cannot rule out the possibility that the released monomers were re-utilized by cells as carbon and energy source.

Cellular debris and some impurities precipitated at pH 1 were further removed by centrifugation. The amount of *R*3HA in the pellets was found to be negligible (≤ 2 wt% of the total released monomers). All kinds of *R*3HA investigated in this study remained soluble. This was rather surprising as carboxylic acids are protonated at low pH and thus show low water solubility. An explanation might be that the

Chapter 3

hydroxy group in β-position contributes to the solubility at low pH. The presence of complexing ions in the supernatant might also prevent precipitation of R3HA. In addition to free R3HA, the solutions contained salts and other cellular components which do not precipitate either at pH 1 or at pH 10.

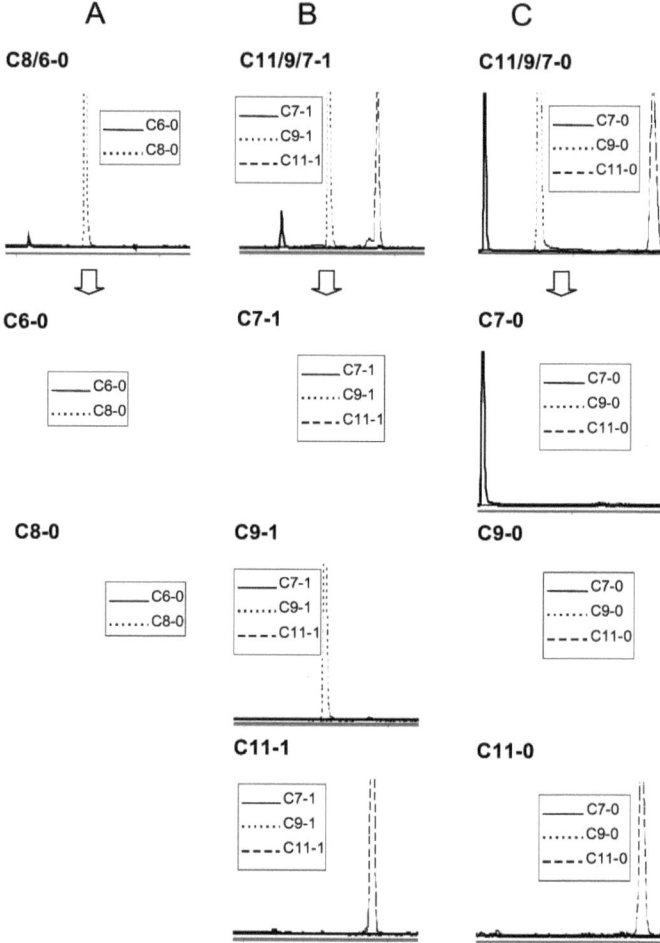

Figure 3.2 Separation of several R3HA analyzed by HPLC. The masses m/z of all the relevant monomers are plotted in each figure as extracted ion chromatogram of the corresponding HPLC-MS spectra. Since measurement was performed with ESI in negative mode, the masses of the corresponding anions could be detected without fragmentation. Panel A: monomers from cells growth on octanoic acid, m/z 131 = 3-hydroxyhexanoate, m/z 159 = 3-hydroxyoctanoate; panel B: monomers from cells growth on 10-undecenoic acid, m/z 143 = 3-hydroxy-6-heptenoate, m/z 171 = 3-hydroxy-8-nonenoate, m/z 199 = 3-hydroxy-10-undecenoate; panel C: monomers from cells growth on undecanoic acid, m/z 145 = 3-hydroxyheptanoate, m/z 173 = 3-hydroxynonanoate, m/z 201 = 3-hydroxyundecanoate. Arrows indicate the monomers before and after separation using column chromatography. All possible R3HA anions are plotted in each picture.

Separation and purification of (R)-3-hydroxycarboxylic acids

Since the obtained monomers from one type of PHA differ in one or more ethylene units of the carbon chain, chemical properties of these compounds were very similar. We were able to separate these monomers by utilizing their unequal hydrophobic properties through column chromatography with a C18 reversed-phase (RP) packing material (for details see materials and methods).

The separation of R3HA was verified by HPLC-MS (Figure 3.2). Since measurements were performed with ESI on negative mode, the masses of the anions corresponding to the distinct R3HA would be detected without fragmentation. Therefore, incomplete separation of R3HA could have been discovered by detection of the corresponding masses of the anions. In Figure 3.2, panels A, B and C represent spectra of monomers obtained from cells grown on octanoic acid, 10-undecenoic acid and undecanoic acid, respectively. R3HA, being produced by *in vivo* degradation of PHA, appear as a mixture of monomers. Pure products were subsequently obtained through the purification process. In this study, we were able to purify R3HA monomers in mg to g quantities.

During the sample collection, fractions with low product concentration were discarded to reduce the working volume for the following purification steps. This explains why a yield of only ~80 wt% for R3HA C8-0 was reached (Table 3.3). The isolation of other R3HA gave similar yields (data not shown). Further purification was carried out by extracting the respective R3HA from combined fractions with tert-butylmethyl ether. This step was accomplished with high yields, e.g., ~98 wt% based on monomer content after column chromatography and 78 wt% based on total released monomer (Table 3.3). After extraction, small amounts of R3HA, e.g., less than 1 µg·mL^{-1} 3-hydroxyoctanoic acid, were detected in the aqueous phase.

Table 3.3 Yields of consecutive isolation steps. The concentrations of 3-hydroxyoctanoic acid after the corresponding isolation step were analyzed by gas chromatography. The concentrations were calculated with respect to 2-ethyl-2-hydroxybutyric acid as internal standard and (R,S)-3-hydroxyoctanoic acid for calibration. Data originated from 4 independent measurements. *The monomer concentration measured in the supernatant after *in vivo* PHA degradation was considered as 100%.

Purification step	C8-0 [mg L^{-1}]	Yield [%]
1. monomer released in supernatant	436 ± 16	100*
2. after column separation	345 ± 2	79 ± 0.4
3. after solvent extraction	341 ± 2	78 ± 0.4

Previously, several R3HA have been separated by using fractional distillation of the corresponding R3HA methyl esters and their subsequent saponification.[92] This process is quite complicated because the free acids had to be derivatized in an additional step. Sample preparation took more than 3 days and distillation itself took 7 days. In addition, large amounts of organic solvents had to be used. The separation process described in this paper (including column chromatography and solvent extraction) involved only a few steps under mild conditions, therefore handling was much easier and could be accomplished within two days. Furthermore, less technical equipment, energy, and organic solvent were used. The organic solvent applied in this study could be recycled. The procedure developed here can easily be transferred to a larger scale.

Chapter 3

Characterization of the purified *(R)*-3-hydroxycarboxylic acids

To determine purities and structures of the produced R3HA, the isolated products were analyzed by ^{1}H NMR spectroscopy (Figure 3.3). Ratios of the respective peak integrals corresponded well with the chemical structures and no side products were detected. Purities of all R3HA were in a similar range and at least higher than 95 wt% as confirmed by both ^{1}H NMR and GC (data not shown). The configuration of the purified monomers was examined by chiral GC, as illustrated for C8-0 in Figure 3.4.

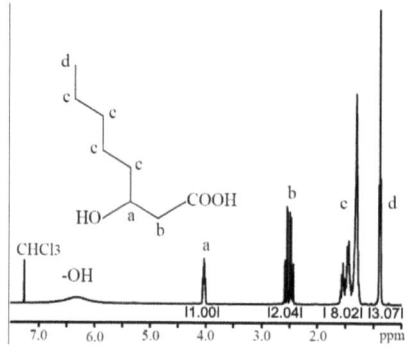

Figure 3.3 Proton (^{1}H) NMR spectrum of *(R)*-3-hydroxyoctanoic acid.

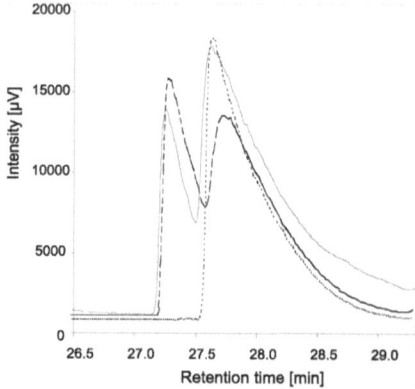

Figure 3.4 Chiral gas chromatography analysis of purified *(R)*-3-hydroxyoctanoic acid (C8-0). Dashed line: racemic *(R,S)*-3-hydroxyoctanoic acid methyl esters; dotted line: 3-hydroxyoctanoic acid methyl ester prepared from C8-0 purified in this study; solid line: mixture [1/1 vol%] of the racemic standard (dashed line) and purified *(R)*-3-hydroxyoctanoic acid methyl ester (dotted line).

The racemic standard *(R,S)*-3-hydroxyoctanoate methyl ester showed two peaks with the retention times (t_R) of 27.3 and 27.7 minutes which corresponded to the two enantiomers. As previously reported, the later peak can be assigned to the *(R)*-enantiomer.[264] Only one peak was detected with the prepared 3-hydroxyoctanoate methyl ester from this study, revealing the high enantiomerical purity of this substance. By mixing 3-hydroxyoctanoate methyl ester and the racemic standard the area of the later peak at 27.7 min increased, confirming the absolute (R)-configuration of purified R3HA. Similar results were obtained for C6-0, C7-0, C9-0, and C11-0. There was no standard material available for the R3HAs with terminal double bonds. Thus, in vivo depolymerization and subsequent isolation is a satisfactory approach to produce enantiomerically pure R3HA.

The melting points (T_m) and enthalpies of the isolated R3HA were analyzed by DSC measurements (Table 3.4). The melting points measured for C8-0, C9-1, C9-0, C11-1, and C11-0 were 22°C, 12°C, 53°C, 39°C and 60°C, respectively. For example, C8-0 seemed to have a less crystalline, i.e., a less ordered solid-state structure than its homologues with longer carbon chain. The weaker intermolecular interactions require less energy to be overcome which results in a lower melting temperature of C8-0. Melting points of R3HA with chain lengths shorter than 8 carbon atoms were below 0°C. Melting points of some non-hydroxylated counterparts are listed for comparison. Enthalpies of C8-0, C9-1, C9-0, C11-1, and C11-0 were found to be between -199 J·g^{-1} and -96 J·g^{-1}.

Table 3.4 Melting points and enthalpies of R3HA measured by DSC. Data originated from at least 2 independent measurements. T_m values of non-hydroxylated counterparts were obtained from the Sigma-Aldrich catalog. *not found in the literature.

R3HA	T_m [°C]	ΔH_f [J g^{-1}]	Non-hydroxylated counterparts	T_m [°C]
C8-0	22 ± 2	-96 ± 11	octanoic acid	16
C9-1	12 ± 2	-199 ± 20	8-nonenoic acid	*
C9-0	53 ± 1	-159 ± 1	nonanoic acid	12
C11-1	39 ± 3	-110 ± 9	10-undecenoic acid	26
C11-0	60 ± 1	-108 ± 5	undecanoic acid	30

Antimicrobial activity of purified R3HA

Previously, it was reported that aromatic (R)-3-hydroxy-n-phenylalkanoic acids have antimicrobial activity against food pathogens such as *Listeria monocytogenes*.[224] In this study, we tested whether some of the purified R3HA also have antibacterial effect on different *Listeria* (*L. innocua*, *L. monocytogenes*, *L. ivanovii*) and other bacterial species (*E. coli* MG1655, *Sal. enterica*, and *Sta. aureus* RN4220). When C8-0, C9-1 and C11-1 were used, the minimal inhibitory concentration (MIC) for all tested *Listeria* species and *Sta. aureus* RN4220 ranged between 1 mM and 5 mM. For *E. coli* and *Sal. enterica* no growth inhibition was detected at investigated concentrations (up to 10 mM for all tested R3HA). As comparison, the commercially available racemic mixture (R/S)-3-hydroxyoctanoic acid and octanoic acid were also investigated for their antibacterial activity. It was observed that the former had lower inhibitory effect (MIC above 10 mM on all tested strains) than its *R*-enantiomer, and the latter did not cause any effect on the bacterial growth of the different species unless the concentration was higher than 50 mM. It suggests that the antibacterial activity is mainly caused by the *R*-enantiomer.

The results that C8-0, C9-1 and C11-1 inhibited the growth of *Listeria* species and *Sta. aureus* with a MIC of 1-5 mM were similar to what was reported recently that (R)-3-hydroxy-phenylalkanoates could inhibit bacterial growth with a MIC of 3-6 mM.[224] These data suggest that not only aromatic but also aliphatic *(R)*-3-hydroxycarboxylic acids are effective against *Listeria* species. Thus, the work described here opens a new route for the preparation of various enantiomerically pure *(R)*-hydroxycarboxylic acids for antimicrobial applications.

Conclusions

In this study, we have demonstrated that various enantiomerically pure *(R)*-3-hydroxycarboxylic acids can be produced and purified via in vivo depolymerization, column chromatography and solvent extraction. Preparative liquid column chromatography with reversed-phase material is proven to be an efficient method for good separation and high recoveries of *R*3HA. Scale-up of the separation process might be accomplished if larger columns and automated systems could be used. Compared to previously described methods,[80, 92] the procedure developed here leads to effective cost reduction, easy downstream processing and an environmentally friendly approach. The procedure is not restricted to the *R*3HA tested in this study. It can probably be applied to other *R*-hydroxyalkanoic acids which are accumulated in bacterial PHA, i.e., approximately 140 potential candidates.

Acknowledgement

We thank Ernst Pletscher for producing the polymer PHUA. We thank Elisabeth Michel for helping with GC analyses. Thanks are given to Roland Hany for NMR measurements, Franziska Amman and Manfred Schmid for DSC measurements. Thanks are also due to Veronika Meyer for careful reading of the manuscript. We greatly acknowledge the financial support by Empa.

Chapter 4

Process engineering for production of chiral hydroxycarboxylic acids from bacterial polyhydroxyalkanoates

Qun Ren, Katinka Ruth, Linda Thöny-Meyer, Manfred Zinn. 2007. Macromolecular Rapid Communications, 28 (22), 2131-2136

Katinka Ruth has carried out the following parts: PHA synthesis, analysis of PHA, purification of monomers.

Chapter 4

Abstract

An efficient process for production of (R)-hydroxycarboxylic acids (RHAs) from polyhydroxyalkanoates (PHA) was developed (Fig. 4.1). It involved the synthesis of PHA in bacteria, followed by bringing the culture broth directly to a pH optimal for *in vivo* PHA degradation, thus avoiding cell collection by centrifugation and pellet resuspension. The optimal pH was maintained to allow maximal release of RHAs. Using this process, cells having a dry weight (w) of 1.8 g L^{-1} and 45% (w/w) PHA exhibited a exponential PHA degradation with initial rate of about 0.059 g L^{-1} h^{-1} in the first 9 hours. Concomitantly, RHAs were released with a rate of 0.057 g L^{-1} h^{-1}. Further incubation of up to 15 hours resulted in almost 90% (w/w) degradation of PHA. Based on this approach in combination with chemostat and a plug flow reactor a continuous process for production of RHAs could be achieved.

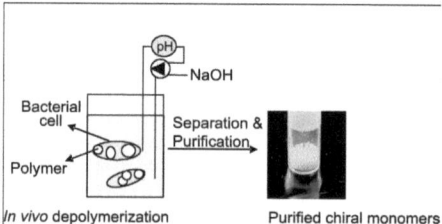

Figure 4.1 This study presents a continuous bioprocess to obtain chiral monomers ((R)-hydroxycarboxylic acids) involving cultivation of *P. putida* GPo1 and *in vivo* depolymerization of its intracellularly accumulated PHA.

Introduction

In the past decades enantiomerically pure compounds have become key synthons for the pharmaceutical industry. Due to growing awareness of the importance of chirality in conjunction with biological activity and consequent pressure from regulatory agencies, the production of racemic mixtures as a starting material for multistep-synthesis has ceased to be a rational commercial option. Many drugs are now synthesized using chiral synthons provided either by kinetic resolution of racemates, asymmetric synthesis or via the naturally occurring chiral pool.[268]

(R)-hydroxycarboxylic acids (RHAs) can be widely used as chiral precursors for several reasons: (i) they contain at least two functional groups: a hydroxyl-group and a carboxyl-group; (ii) the functional groups are convenient to modify; and (iii) a new chiral center can be introduced. The reported compounds using RHAs as chiral building blocks comprise the macrocyclic component of the antibiotic elaiophylin,[262] the hydroxyacyl hydrazines in visconsin, a peptide antibiotic,[131] pharmaceuticals such as captopril and β-lactams,[81, 269] and fungicides such as norpyrenophorin and vermiculin.[261]

It has been reported that RHAs can be obtained by hydrolysis of biotechnologically synthesized polyhydroxyalkanoates (PHA). PHA are microbial polyesters, which are accumulated as a carbon and energy reserve under particular environmental conditions, i.e., under carbon limitation.[252] Due to the asymmetric carbon atom in the β position, PHA are optically active polymers containing only the R-enantiomer. More than 140 different monomers containing different functional groups have been reported to be incorporated into PHA.[13, 78] Chemical degradation of PHA has been applied to generate RHAs.[92, 93] The limitations of this method are that organic solvents were used in large amounts, multi-steps (purification of PHA, acidic hydrolysis of PHA, distillation and saponification) had to be applied, and the production efficiency was rather low.

Recently, it was reported that RHAs could also be produced from PHA by *in vivo* depolymerization,[70, 264] a process where PHA biosynthesis and subsequent degradation of accumulated PHA take place inside the cells. To do so, cells containing PHA were resuspended in suitable solutions having the optimal pH for RHA production.[70, 264] RHAs produced in such way are excreted into the medium. However, the scale-up of these processes is difficult, because the released RHAs are acidic and continuously change the extracellular pH, which leads to unfavorable conditions for PHA degradation. Thus, it was necessary to develop a process wherein the pH can be regulated to allow maximal release of RHAs. Furthermore, in previously reported methods cells containing PHA were collected first by centrifugation or filtration, and then resuspended. The released RHAs needed to be further separated from the cells by a second step of centrifugation or filtration. The whole process was time and energy consuming and also resulted in loss of cell materials. The present work describes a process for RHA production from bacterial PHA, which overcomes the limitations of previous methods.

Experimental Part

Bacterial strains and cultivation conditions

The wild-type strain *Pseudomonas putida* GPo1 (ATCC 29347) was used throughout the entire study. In batch culture, cells were grown in modified E2 medium supplemented with 10 mM sodium citrate (Fluka, Switzerland),[31, 53, 63] and incubated in shaking flasks at 30°C and 160 rpm. They were used as inocula for a batch culture that was then changed to continuous cultivation. The continuous cultivation was performed in a reactor with a working volume of 1.8 L as previously described.[31, 63] Octanoic acid (Fluka, Switzerland) of 15 mM was used as the sole carbon source. A dilution rate (D) of 0.1 h^{-1} and a carbon to nitrogen ratio (C/N) of 15 (g g^{-1}) were applied in the medium feed.[31, 63] PHA containing *R*-3-hydroxyoctanoic acid (R3HO) and *R*-3-hydroxyhexanoic acid (R3HH) was produced and tested for release of RHA monomers. If necessary, a plug flow (PF) reactor was used. The PF reactor consisted of a silicon tube (10 mm inner diameter) with lengths selected to give the desired residence time of 3, 6 and 9 hours.

Production of RHAs

Culture broth harvested at the steady state was brought to an alkaline pH (8.0 to 11.5) by adding 2N NaOH. It was incubated for different time periods as described in Results and Discussion. Subsequently, the suspension was acidified to pH 1 with concentrated HCl.[270] The supernatant which contained RHA monomers was obtained by centrifugation (4,500 x g, 4°C, 20 min; Multifuge 3 S-R, Osterode, Germany). The supernatant was filtered once with filter papers (Macherey Nagel, Switzerland)

and once with membrane filters (Regenerated cellulose membranes, pore size 0.45 µm, Whatman, Switzerland). The filtered solution was subjected to further purification. If necessary, the pH of the culture broth was maintained constant using a pH controller (Instrumentation Laboratory, Switzerland) by automatic titration with 2N NaOH. The deviation of the pH value from the set point was ± 0.3.

Analysis of RHAs and intracellular PHA

For RHA analysis, 8 mL of RHA-containing solution or 5-15 mg isolated RHAs were filled into a 10 mL Pyrex tube and treated according to the method described previously.[250, 270] Samples were subsequently analyzed on a GC (Fisons Instruments, Rodano, Italy) equipped with a polar fused silica capillary column (Supelcowax-10: length 30 m; inside diameter 0.31 mm; film thickness 0.5 µm; Supelco, Buchs, Switzerland).[270] For the analysis of intracellular PHA, the culture broth was centrifuged (10,000 x g; 4°C; 15 min), and the cell pellet was lyophilized for 48 hours. Around 50 mg of dry biomass were filled into a 10 mL Pyrex tube, and PHA content and monomer composition were analyzed as previously described.[270]

Separation and purification of RHA

Separation and purification of different monomers was performed by C18-reversed phase chromatography using a glass column packed with silica gel 100 (particle size 0.040-0.063 mm; max. surface coverage 17-18% C; Fluka), followed by subsequent solvent extraction with tert-butyl methyl ether.[270] The purified RHAs were analyzed by HPLC-MS and GC for their presence and purity.[270]

Chirality analysis of monomers

Commercial R,S-3-hydroxyoctanoic acid (Sigma-Aldrich, Seelze, Germany) and the monomers purified in this study were methylated.[256] Separation of the enantiomers was performed on a GC equipped with a Beta-DEX 120 column (fused silica capillary column; length 30 m; inside diameter 0.25 mm; film thickness 0.25 µm; Supelco, Buchs, Switzerland) according to a method described previously.[270]

^{1}H NMR

Proton nuclear magnetic resonance experiments in $CDCl_3$ solution were performed on a Bruker AV-400 spectrometer. Chemical shifts were given in ppm relative to the signal of deuterated chloroform as internal reference (^{1}H NMR: 7.26 ppm).

Reproducibility

In this study, the absolute values regarding PHA degradation and RHA production varied due to different cell dry weights and intracellular PHA contents for different batches of cultivation; however, the *in vivo* degradation of PHA was influenced by the environmental conditions in the same manner in all experiments. For the same batch of samples the experiments were performed in duplicates. The standard deviation (SD) of the data obtained from GC and HPLC-MS analysis was ±5% and ±10%, respectively.

Results and Discussion

Preparation of RHAs without controlling the pH of the culture broth

In all previously reported *in vivo* depolymerization methods cells containing PHA have been first collected by centrifugation or filtration and then resuspended in suitable solutions having an optimal pH for RHA production.[70, 264] In this study, we tested whether RHAs could be released when the cell broth was directly brought to the optimal pH. *P. putida* GPo1 was cultivated in a continuous culture. PHA containing R3HO and R3HH was produced to about 45% (weight/weight (w/w)) of the cell dry weight (CDW) (1.8 gL^{-1}) at the steady state. The culture broth was collected and then directly adjusted to different pH (8.0, 8.5, 9.0, 9.5, 10.0, 10.5, 11.0 and 11.5) to investigate PHA degradation and RHA release (data not shown). The optimal pH tested here was found to be between 9.5 and 10.5, similar to what we observed before.[264] With the initial pH of 10.0 at room temperature an exponential PHA degradation pattern was observed (Fig. 4.2A). PHA was degraded from 0.77 gL^{-1} to 0.31 gL^{-1} in 12 hours, which is about 60% of the originally accumulated PHA. Correspondingly, the amount of RHAs in the supernatant was increased from 0 to about 0.43 gL^{-1} in 12 hours. Further incubation up to 24 hours did not lead to significantly better yield (Fig. 4.2A).

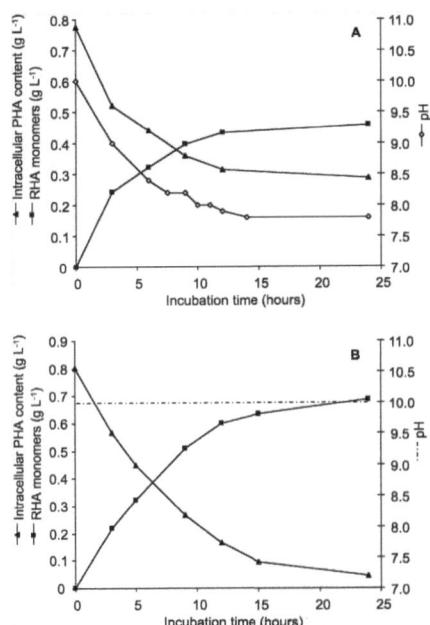

Figure 4.2 Preparation of RHA monomers via *in vivo* degradation of bacterial PHA. *P. putida* GPo1 cells were cultivated in a continuous culture and the culture broth was collected after reaching the steady state. The pH of the cell broth was adjusted to 10.0 by adding 2 N NaOH, then either was not further controlled (A) or was controlled at 10.0 (B). Degradation of PHA and release of RHAs under both conditions was followed for 24 h.

It was also observed that the pH of the culture broth dropped in a similar manner to that of PHA degradation during the incubation, namely from an initial pH of 10.0 to about 8.0 after 12 hours (Fig. 4.2A). This suggested that the release of monomeric acids constantly lowered the environmental pH of the culture broth, resulting in less favorable conditions for PHA degradation.

Preparation of RHAs by controlling the pH of the culture broth

To investigate whether a better yield of RHA production could be achieved by keeping the pH of the cell broth constant at a level optimal for PHA degradation, the culture broth of a continuous culture was collected, and adjusted to and maintained at pH 10.0 by a pH controller. The degradation of PHA and the release of RHAs were followed for 24 hours at room temperature (Fig. 4.2B). A linear PHA degradation pattern was observed in the first 9 hours, i.e., PHA degraded from 0.80 gL^{-1} to 0.27 gL^{-1}, resulting in a degradation rate of 0.074 gL^{-1}h^{-1}. Correspondingly, R3HO and R3HH were produced with a rate of 0.057 gL^{-1}h^{-1} and reached 0.51 gL^{-1} in 9 hours. After 15 hours almost 90% (w/w) of the total PHA was degraded. Concomitantly, about 0.63 gL^{-1} RHAs were found to be released into the supernatant, leading to a productivity of 0.042 gL^{-1}h^{-1}. It has been reported previously that in *P. putida* GPo1 at most 9.7% of PHA containing R3HO and R3HH could be depolymerized after 4 days.[70] By contrast, the obtained yield in this study was almost one order of magnitude higher and the efficiency was also much higher. Recently, we have shown that the efficiency of PHA degradation reached 90% (w/w) after 9-12 hours at 30°C when cells were collected by centrifugation and resuspended in a phosphate buffer with a pH of 10-11.[264]

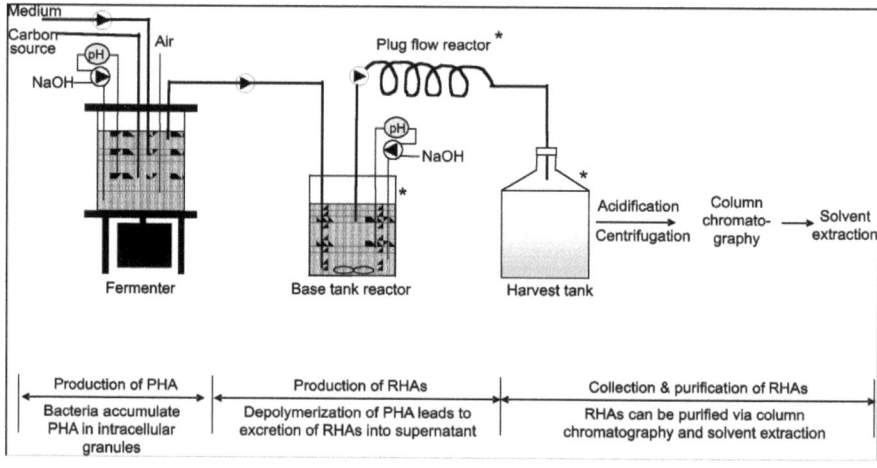

Figure 4.3 Schematic representation of a continuous process for production of RHAs out of bacterial PHA. *Open system where no sterilization is needed.

Considering that the efficiency of PHA degradation is in the same range, and that previously reported method is not suitable for high PHA concentrations because the buffer strength may not be sufficient to prevent the constant acidification caused by the released monomeric acids [270], the method developed here using controlled pH is a real improvement. Furthermore, the new method avoided centrifugation and resuspension of cells, and consequently saves time and energy.

Continuous preparation of RHAs by *in vivo* degradation of PHA

We further designed a process for continuous RHA preparation using the method developed in this study in combination with chemostat (Fig. 4.3). The culture broth of a chemostat was continuously directed to a stirred tank reactor where the pH value was adjusted to and controlled at 10.0 by adding 2 N NaOH (Fig. 4.3). The influence of the average residence time of the cells in the tank reactor on PHA degradation was tested: 5, 7.5, 10 and 15 hours (Fig. 4.4A). It was found that PHA degraded from 0.8 to 0.3 gL^{-1}, which is about 60% of total PHA, with the residence time of 10 hours at room temperature. Prolonged residence time of up to 15 hours did not lead to a significantly improved yield.

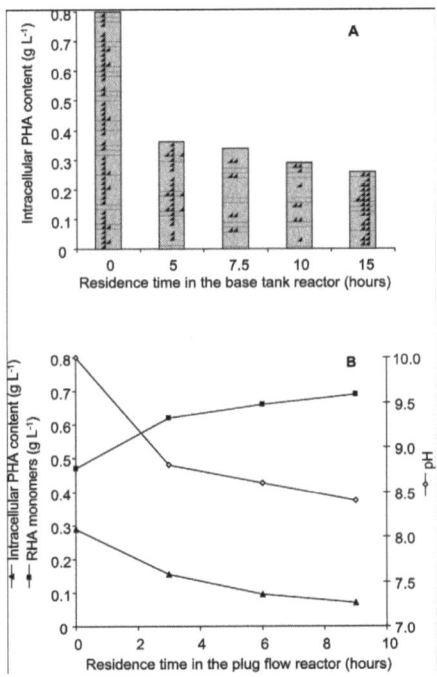

Figure 4.4 Continuous preparation of RHAs through a PF tube reactor in combination with chemostat. Cells were cultivated in a continuous culture and directed to a stirred base tank reactor after reaching the steady state. The pH of the tank reactor was kept at 10.0 by automatic titration of 2N NaOH with a pH controller. (A) Influence of the residence time of the in the tank reactor on PHA degradation. (B) Cells were incubated in the tank reactor for 10 h, and were further delivered to a PF tube reactor. The residence time of the cells in the PF tube, adjusted by the tube length, was studied regarding further degradation of PHA and release of RHA monomers.

To enhance PHA degradation and RHA production, a plug flow (PF) reactor was applied (Fig. 4.3). Culture broth was delivered from the tank reactor to the PF tube by a pump. The pumping flow rate was the same as the flow rate of the chemostat. After the cells had an average residence time of 10 hours in the tank reactor with a set pH of 10, they further degraded PHA and released RHAs inside the PF tube (Fig. 4.4B). After 6 hours in the PF reactor almost 90% (w/w) of the originally accumulated PHA was degraded and 0.66 g RHAs L^{-1} could be obtained, resulting in a productivity of 0.041 $gL^{-1}h^{-1}$ (Fig. 4.4B). This productivity is similar to that achieved by the batch preparation of RHAs. However, when chemostat is used a PF reactor might be a more efficient means for production of RHAs since it avoids time delays in cell collection from chemostat reactor, change of equipment and readjustment of parameters, thus enables a continuous production process.

Purification and characterization of produced RHAs

To test whether the RHAs obtained via the process developed here could also be isolated by the method we reported previously,[270] purification of R3HO and R3HH was performed. It was found that both monomers could be purified with an overall yield of about 80% (w/w) relevant to the total originally released R3HO and R3HH in the supernatant. The purified 3HO was further analyzed for its purity and structure by ^{1}H NMR spectroscopy. It was revealed that ratios of the respective peak integrals corresponded well with the chemical structure and no side products were detected (data not shown). The purity of the obtained 3HO was higher than 95% (w/w) based on both GC and NMR analysis. The configuration of the purified 3HO was further examined by chiral GC (Fig. 4.5).[270] The racemic *(R,S)*-3-hydroxyoctanoate methyl ester was used as reference. Only one peak was detected with the prepared 3HO methyl ester (Fig. 4.5), which revealed the high enantiomerical purity of this substance. When 3HO methyl ester and the racemic standard were mixed, the area of the peak eluting at *R*-enatiomer increased (Fig. 4.5), confirming the absolute *(R)*-configuration of purified 3HO. Thus, the new process developed in this study for *in vivo* depolymerization is a feasible approach to produce enantiomerically pure RHAs.

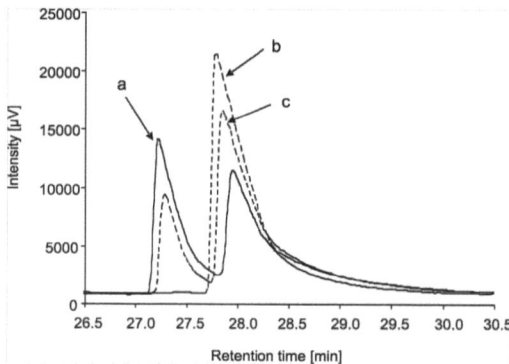

Figure 4.5 Chiral GC analysis of purified (*R*)-3-hydroxyoctanoic acid (R3HO): a, racemic (*R*)/(*S*)-3-hydroxyoctanoic acid methyl esters; b, 3-hydroxyoctanoic acid methyl ester prepared from R3HO purified in this study; c, mixture (1/1 vol.-%) of the racemic reference and purified R3HO methyl esters.

Conclusions

In this report we have developed an efficient and easy-to-handle process to produce enantiomerically pure RHAs from bacterial PHA, which only requires controlling of pH to maintain maximal PHA degradation. The reduction to a single step of centrifugation or filtration is an improvement that should simplify the scale-up of RHA production. Application of PF reactor seems to be an efficient means for continuous RHA production at the laboratory scale. However, the possible obstacles have to be considered such as cell aggregation and cell adhesion inside the PF tube when scale-up is performed. A further improvement of the method may be brought at the step of RHA purification since large volumes have to be handled by chromatography. The improvement of methods for RHA production will certainly facilitate the applications of these chiral compounds as chemical synthons.

Acknowledgements

We thank Roland Hany for NMR measurements, Ernst Pletscher and René Hartmann for assisting with continuous cultivation.

Chapter 5

Degradation of polyhydroxyalkanoates enhances alkaline stress tolerance in *P. putida* GPo1

Katinka Ruth, Thomas Egli, Qun Ren. Manuscript in preparation.

Chapter 5

Abstract

In an alkaline milieu, *Pseudomonas putida* GPo1 degrades intracellular polyhydroxyalkanoates (PHA) and secretes the degradation products, (R)-hydroxycarboxylic acids, into extracellular environment. The physiological role of this action has not yet been investigated. In this study, we observed that release of (R)-hydroxycarboxylic acids decreased the surrounding pH of the GPo1 cells, and the intracellular pH remained below 9.0 for at least one hour when cells were exposed to pH 10.5. Furthermore, *P. putida* GPo1 showed higher survival ability and thus enhanced alkaline stress tolerance compared to its PHA depolymerase deficient counterpart *P. putida* GPo500. Our results suggest that PHA degradation might contribute to cellular processes of pH homeostasis in *P. putida* GPo1.

Introduction

A wide variety of microorganisms can synthesize polyhydroxyalkanoates (PHA) as intracellular storage materials.[21, 78, 97] These biodegradable and biocompatible polyesters are accumulated in the presence of an excess of a carbon source and if growth is limited by another nutrient (e.g., nitrogen).[21, 174] PHA serve as carbon and energy reservoirs.[78, 97] In the absence of a suitable exogenous carbon source or energy source, intracellular PHA are degraded into monomeric units which can be channeled back to cellular catabolism. Synthesis and degradation of PHA are part of a cycle starting from 3-hydroxyacyl-CoA.[40] These precursors are intermediates of β-oxidation and can be converted into PHA by PHA polymerases.[97] Degradation of PHA is carried out by PHA depolymerases.[15, 271] Recently, *in vitro* studies showed that the PHA depolymerase of *Pseudomonas putida* KT2442 hydrolyzes PHA to monomeric acids and that the enzyme behaves like a typical serine hydrolase with an optimal activity under alkaline conditions.[46]

It has been proposed that accumulation and degradation of PHA is one strategy by which bacteria can improve the establishment, proliferation, and survival in competitive settings such as soil.[272] Understanding the role played by PHA under stress conditions is of fundamental importance in microbial ecology and not yet fully understood. Early works suggested an association of PHA utilization with respiration and oxidative phosphorylation in *Ralstonia eutropha* and *Azotobacter beijerinckii* in a manner that PHA can serve as redox regulator.[21, 166, 273] In a recent study, an increasing guanosine tetraphosphate (ppGpp) level appeared to occur concomitantly with PHA degradation.[159] This phenomenon was observed only in wild-type *P. putida* and not in a PHA depolymerase deficient strain unable to degrade the polymer. In *Escherichia coli,* ppGpp can induce, among other factors, the expression of the *rpoS* gene, which encodes a sigma factor involved in regulation of stress tolerance.[274, 275] Furthermore, a decreasing specific growth rate is a stress factor that has been reported to induce RpoS.[276]

In order to further increase our understanding of the role played by PHA in survival and proliferation of *P. putida* GPo1, we conducted experiments in which *P. putida* and its PHA depolymerase-negative mutant were evaluated for metabolic versatility and for the ability to endure various environmental pH. In this study, we demonstrated that *P. putida* can cope with stress caused by alkaline conditions. Even though acidic soil is encoutered more often, alkaline condition can be found in several environments like polluted soils, industrial waste, or soda lakes. Our results suggested that accumulated PHA can contribute to efficient pH homeostasis.

Material and Methods

Cultivation of *Pseudomonas putida*

The wild-type strain *P. putida* GPo1 (ATCC 29347) was cultivated in chemostat cultures at a dilution rate of 0.1 h^{-1} under dual-nutrient-limited growth conditions.[31] Continuous cultivation was performed as previously reported.[264, 265] Release of monomers from PHA was carried out at different pH values (pH 4-11).[264] The intracellular PHA content and PHA monomers in the supernatant were determined by gas chromatographically detection of the corresponding propyl or methyl esters.[250]

Shake flask cultivation of GPo1 and the PHA depolymerase deficient *P. putida* GPo500[17] was performed in 100 mL E2 medium supplemented with 15 mM sodium octanoate as carbon source at 30°C.[62]

pH challenge protocol

Bacteria were grown in shake flasks until the early stationary phase (OD$_{600}$ = 3) in E2 medium supplemented with 15 mM sodium octanoate at 30°C. The cells were harvested by centrifugation (10'000xg) and rinsed twice with saline solution. The pellets were then resuspended in and diluted to 5·10^8 cells per mL with 50 mM potassium phosphate buffer having different pH values. The pH challenge was performed at 30°C in Eppendorf tubes. Aliquots were removed at different time intervals, and the viable bacterial cells were counted immediately after exposure to pH challenge (time zero) and then every 10 min by serial dilutions onto triplicate LB agar plates. The variation coefficient for replicate platings was ≤25%. Survival of wild type GPo1 was set as 100%.

Measurement of pH

Extracellular pH (pH$_{ex}$) was determined by a pH meter 692 with glass electrode (Metrohm, Herisau, Switzerland) calibrated with color-coded buffer standard solutions (Fluka, Buchs, Switzerland). Intracellular pH (pH$_{in}$) was measured using a fluorescence staining method as previously reported.[251] The fluorescence intensity of pH$_{in}$ was measured in a FLx800 Microplate Fluorescence Reader (BioTek instruments, Witec AG, Littau, Switzerland) with 485 nm excitation wave length and 528 nm emission wave length (slit widths 20 nm, sensitivity 60). For calibration, pH$_{ex}$ was measured on same settings after equilibrating pH$_{ex}$ and pH$_{in}$ by addition of the antibiotics valinomycin and nigericin (both 1µM final concentration). These antibiotics destroy pH gradient and membrane potential within microorganisms by irreversibly opening the proton channels. The background was determined by filtering the cell suspension trough a 0.22 µm pore-size membrane filter and measuring the filtrate (extracellular fluorescence).

Chapter 5

Results and Discussion

Extracellular and intracellular pH of *P. putida* cells exposed to different environmental pH

P. putida GPo1 was cultivated with octanoate as the sole carbon source in chemostat culture which led to cells with a cell-dry-weight of ~1.4 gL^{-1} and ~45% (gg^{-1}) PHA at the steady-state. Previously, it has been reported that PHA could be degraded through *in vivo* depolymerization when the cells were exposed to alkaline conditions, and PHA monomers were excreted into the media.[264] In this study, we observed that during *in vivo* depolymerization at alkaline pH the released monomers constantly lowered the environmental pH of the culture broth as shown in Figure 1. When incubated in 50 mM potassium phosphate having an initial pH of 10.0, PHA monomers were released immediately and reached concentrations of ~100 mgL^{-1} (*R*)-3-hydroxyoctanoic acid (C8) in one hour, and extracellular pH dropped to 8.8. Further incubation up to six hours led to a pH value of 8.0 and ~550 mgL^{-1} C8 in the supernatant (Fig. 1). Whereas when initial extracellular pH was neutral (pH 7.0) or acidic (pH 4.0) nearly no change of pH was observed and only low amounts monomers were detected (<50 mgL^{-1} C8 for pH 4; <150 mgL^{-1} C8 for pH 7) (data not shown). Our data suggested that PHA degradation is triggered by alkaline conditions. Since PHA monomers are *R*-3-hydroxycarboxylic acids, PHA degradation may enhance pH homeostasis.

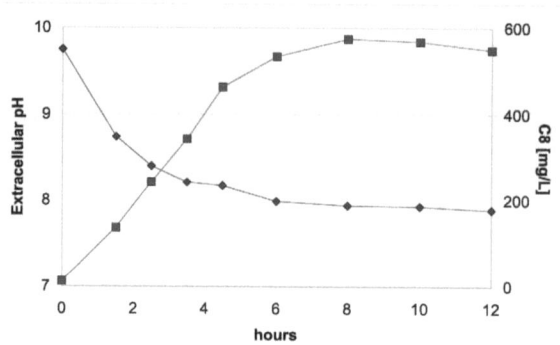

Figure 5.1 Change of extracellular pH (diamond) and release of PHA monomer (*R*)-3-hydroxyoctanoic acid (C8, square) during *in vivo* depolymerization of *P. putida* GPo1. The 50 mM potassium phosphate buffer had an initial pH value of 10. The trend was confirmed by 3 independent measurements.

We also investigated whether the presence of carbon could prevent PHA degradation. In the same experimental settings as described above, 15 mM sodium octanoate was added to phosphate buffer and the pH adjusted to different pH values. It was found that the presence of carbon did not influence monomer excretion within 12 hours and solely the pH value affected monomer release (data not shown). PHA degradation has been proposed to be a central feature of survival physiology when cells are faced with carbon limitation and lack of carbon might direct a regulatory system for transcription of PHA depolymerase (PhaZ).[47] Our data indicated that in *P. putida* GPo1 PHA may function for improved survival ability not only during carbon limitation, but also during environmental pH stress since activity of PhaZ was promoted by alkaline pH, both in presence and absence of carbon.

A significant change in the surrounding pH has a negative impact on cellular metabolism. Cells are equipped by tools such as proton pumps and enzymes to build up a membrane potential which enables maintenance of their cellular milieu. However, harsh conditions can overburden these mechanisms and result in cell lysis. It would be of great interest to estimate the extent up to which *P. putida* GPo1 can bypass harsh condition and regulate its intracellular pH. Therefore, the intracellular pH of GPo1 cultivated in a chemostat was measured when cells were exposed to phosphate buffers adjusted to pH values between 4.0 and 11.0. We found that *P. putida* GPo1 was able to maintain a physiological pH between 6.6 and 8.8 when exposed to buffers with pH values ranging between 6.0 and 10.5 (Fig. 5.2).

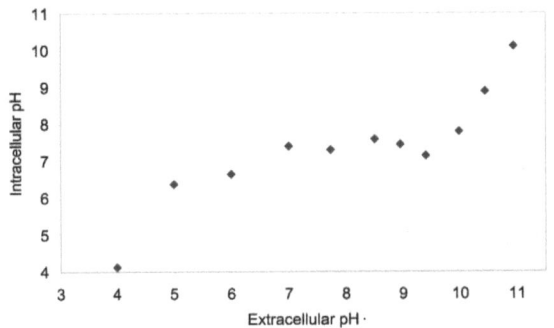

Figure 5.2 Intracellular pH of *P. putida* GPo1 with respect to extracellular pH. Fluorescence-stained cells were exposed to the buffers with their distinct pH for 15 minutes. Standard deviation was ± 0.3 pH units. The trend was confirmed by 3 independent measurements.

P. putida GPo1 seems to be able to maintain its intracellular pH and under alkaline conditions it might be accomplished by PHA degradation since the extracellular pH was decreased by released PHA monomeric acids (Fig. 5.1), which could also neutralize the intracellular milieu. With an extracellular milieu of pH 10.5, we observed that cells kept the pH below 9.0 for at least 1 hour.

PHA degradation enhances cell survival under alkaline conditions

Early-stationary phase cells of *P. putida* GPo1 and *P. putida* GPo500, cultivated under PHA-accumulating conditions, were challenged by alkaline pH. It was found that GPo1 was more resistant than the PHA depolymerase-negative counterpart GPo500, as measured by viability counting. When bacteria were incubated in 50 mM potassium phosphate having a pH of 10.5, the mutant GPo500 cells died rapidly. After 20 min of exposure, only 27% of viable cells were found for GPo500 compared with viability of GPo1 (100%). At neutral pH (pH 7) viability of both strains, GPo1 and GPo500 were similar. Our results revealed that the ability of *P. putida* to tolerate high pH was significantly affected by the lack of PHA degradation, indicating that PHA plays an important role in the alkaline stress response in *P. putida*. Recently, it was reported that guanosine tetraphosphate (ppGpp) concentration was increased simultaneously with PHA degradation.[159] This phenomenon was observed only in wild-type *P. putida* and not in a PHA depolymerase deficient strain unable to degrade the polymer. In *Escherichia coli*, ppGpp induces expression of *rpoS*, which encodes a sigma factor involved in regulation of stress tolerance.[275] RpoS activates a cellular stress response to provide cells with the ability to survive the actual stress.[277] It has been shown that one of the promoters that control PHA synthesis in *Azotobacter vinelandii* is regulated by RpoS.[278] It is possible that PHA depolymerase (PhaZ) activity might be increased under alkaline stress through regulation of RpoS, thus acidic PHA monomers could be produced to cope with the high environmental pH.

Hypothesis for pH homeostasis in *P. putida* cells

Up to now, not much is known about the influence of PHA accumulation or degradation with respect to pH homeostasis. *In vivo*, (R)-hydroxycarboxylic acids are produced by hydrolysis of PHA, catalyzed by PhaZ.[46] PhaZ of *P. putida* KT2442 has an optimum at pH 8.8 and it still has 50% activity at pH 10.5.[46] The pool of (R)-hydroxycarboxylic acids can be further channeled back to cellular catabolism, back to PHA or secreted into culture media. Based on the data obtained in this study, and knowledge gained previously (Fig. 5.3) we propose a hypothesis how cells cope with environmental alkaline pH through PHA degradation: The higher the extracellular pH is, the higher the intracellular pH gets in the course of time. This increases the activity of PHA depolymerase since hydrolysis is favored in alkaline milieu,[46] leading to higher concentrations of (R)-hydroxycarboxylic acids secreted into the culture media, thereby enhancing pH homeostasis.

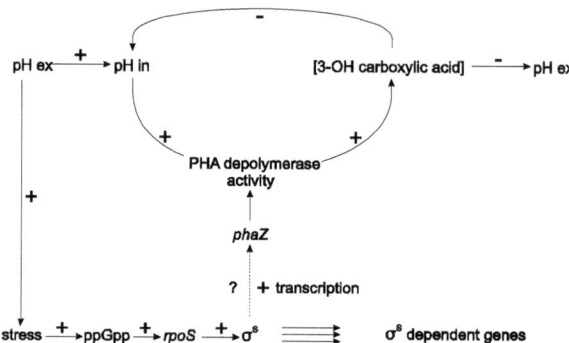

Figure 5.3 Hypothesis of alkaline stress tolerance in *P. putida* GPo1 through PHA degradation. (+) increase; (-) reduction. Lower part adapted from Hegge-Adonis.[277]

It is also possible that alkaline pH triggers induction of stress factors such as *rpoS* via ppGpp. *rpoS* might enhance activity of PhaZ as part of a general stress response. Hence, genetic regulation of PHA metabolism might also play a role in pH homeostasis. Further experiments such as examination of ppGpp content and *phaZ* expression at different environmental pH are required to study the involvement of PHA degradation under alkaline stress.

Conclusion

We studied the potential role of PHA in pH homeostasis in *P. putida*. PHA may be involved in a protection mechanism of *Pseudomonas* in alkaline milieu. Whether or not the medium contained a carbon source did not affect the observed trends of PHA degradation. Thus, the ability to accumulate and degrade PHA might serve to survive stress not only caused by limitation of nutrients, but also caused by alkaline conditions. We propose that the pH might have a direct influence on the activity of PHA depolymerase, and / or the activity of PHA depolymerase could be regulated by *rpoS*, which is produced under stress conditions (Fig. 5.3). Both actions could result in increased concentrations of (*R*)-hydroxycarboxylic acids, which, in turn, would lower the pH and enhance the tolerance of the alkaline stress (Fig. 5.3).

Chapter 6

Identification of two acyl-CoA synthetases
from *Pseudomonas putida* GPo1:
One is located at the surface of
polyhydroxyalkanoates (PHA) granules

Katinka Ruth, Guy de Roo, Thomas Egli, Qun Ren. 2008. Biomacromolecules, 9 (6), 1652-1659.

Abstract

Pseudomonas putida GPo1 is able to accumulate polyhydroxyalkanoates (PHA) in the form of intracellular granules as storage materials. PHA granules were isolated and analyzed for protein activities. An acyl-CoA-synthetase (ACS1) activity was detected from the purified PHA granules. The corresponding gene *acs1* was then cloned from *P. putida* GPo1. With the genomic walking technique, a homologue *acs2* located upstream of *acs1* was discovered and cloned. Fusions of both *acs1* and *acs2* with the gene encoding the green fluorescent protein (GFP) were constructed and expressed in GPo1. *In vivo* fluorescence microscopy studies showed that the fluorescence generated from the ACS1-GFP was mainly associated with the PHA granules, whereas that from ACS2-GFP was mainly with the membrane of the cells. In the control strain (containing GFP alone) fluorescence was distributed evenly in the cytoplasm. We concluded that ACS1 is located on the PHA granules and may play a central role in mobilization of PHA, e.g., conversion of hydroxycarboxylic acid monomers to hydroxycarboxyl-CoA, which can be further utilized by the cells.

Introduction

Polyhydroxyalkanoates (PHA) are naturally occurring polyesters that are produced by a wide variety of microorganisms as carbon and energy storage or reducing power.[21, 174] They can be classified into three groups based on the number of carbon atoms in the monomer units:[13, 174] short-chain-length (scl) PHA, which consist of 3-5 carbon atoms, medium-chain-length (mcl) PHA, containing 6-14 carbon atoms, and long-chain-length (lcl) PHA with more than 14 carbon atoms. PHA are normally synthesized when cells are cultured in the presence of an excess carbon source and when growth is limited by the lack of an essential nutrient.[13, 166] If the cells are under carbon limitation, the accumulated PHA can be degraded to monomers which can be re-utilized by the bacteria as a carbon and energy source.[21] Currently, it is not clear how the physiological conditions trigger mcl-PHA accumulation and degradation.

In vivo, PHA are accumulated in intracellular granules which are covered by a surface layer composed of proteins and phospholipids.[198, 279] The proteins consist of structural proteins (so called phasins), catalytic proteins, and regulatory proteins such as ApdA (activator of polymer degradation A).[280] Phasins are in general low molecular weight proteins and are assumed to form a protein layer at the surface of the granules.[198] They are thought to provide an interface between the hydrophilic cytoplasm and the hydrophobic core of the PHA inclusion and in this way to prevent their coalescence.[281] Studies with phasins such as PhaP, PhaF and PhaI suggest that phasins play an important, but not essential role in PHA metabolism.[282] The catalytic proteins identified up to now are mainly PHA polymerases and depolymerases, which are involved in the last step of PHA synthesis and the first step of PHA degradation, respectively. Previously, it has been reported that the PHA depolymerase degrades the polymer into hydroxycarboxylic acids which can be released into the cytoplasm and subsequently excreted by cells into the environment.[46, 80, 264] In order to recycle the monomeric acids, cells have to activate them to a CoA-linked form by enzymes such as acyl-CoA synthetase. Since the diffusion of monomeric acids away from the PHA granule into the cytoplasm may lead to a cytoplasmic pH change, it would be more efficient and practical if the enzyme for activating the monomeric acid is also located on the PHA granule surface.

Acyl-CoA synthetases (ACS) are a ubiquitous family of enzymes that activate fatty acids by ligating CoA to their carboxy residues.[232-234] Most of them are involved in the β-oxidation fatty acid degradation pathways. One of the most extensively studied ACSs is the *Escherichia coli* FadD.[239, 240] *E. coli* FadD is a cytoplasmic membrane-associated protein, and it activates exogenous long-chain fatty acids into metabolically active CoA thioesters when they are transported across the cytoplasmic membrane.[239, 241] In *P. putida* U two FadD homologues have been reported: FadD1 and FadD2.[38, 244] The *fadD1*-deficient mutant was not able to grow on fatty acids with acyl chains longer than C_4 as sole carbon sources.[248] However, prolonged incubation of 80 hours could restore growth.[248] Disruption of the *fadD2* gene did not have any effect on the catabolism of fatty acids.[248] It was concluded that FadD1 is an enzyme involved in the physiological degradation of fatty acids, and FadD2 is induced when FadD1 is inactivated. Once adapted, *fadD1* mutants of *P. putida* U were able to resume growth and PHA synthesis, reaching similar biomass and PHA contents to the parental strain.[248] The produced PHA had similar morphology and monomer compositions to the parental strain.[248]

In this study, we identified two acyl-CoA-synthetases in *P. putida* GPo1. We found that one of them (ACS1) was associated with PHA granules. This ACS seems to have a central role in mobilizing mcl-PHA. Based on our data and the results obtained previously, a model is proposed for PHA metabolism at the enzymatic level.

Materials and Methods

Chemicals and reagents

All chemicals were purchased from Sigma-Aldrich (Buchs, Switzerland) and of analytical grade. (R)-3-hydroxycarboxylic acids (C_5-C_{16}) were produced as described by de Roo et al.[92] Restriction enzymes, Taq DNA polymerase, and nucleotides were supplied by Bioconcept (Allschwil, Switzerland). DNA isolation kits were obtained from Biorad (Reinach, Switzerland), gel extraction kits were from Qiagen (Hombrechtikon, Switzerland), and plasmid isolation kits were purchased from Sigma-Aldrich (Buchs, Switzerland), the GenomeWalker Universal kit was from Clontech (California, USA).

Bacterial strains, plasmids, culture media, and growth

Bacterial strains and plasmids used in this study are listed in Table 6.1. Recombinants of *E. coli* DH5α transformed with various plasmids were grown at 37°C in tryptic soy broth (TSB) (Sigma) or on tryptic soy agar plates (Sigma). *E. coli* JMU194 was cultivated in 0.1 NE2 medium[179] supplemented with 0.2% yeast extract and 2 mM hexadecanoate.[191] *P. putida* GPo1 was either grown in TSB, or in E2-medium[62] supplemented with either 25 mM glucose or 15 mM sodium octanoate or 2 mM octadecanoic acid (dissolved in 0.5% Brij56) as the sole carbon source. All cultures were performed in shake flasks at 30°C and 150 rpm. Growth was measured at OD_{600} with a Digitana spectrophotometer

(Horgen, Switzerland). If necessary, ampicilin (100 µg mL^{-1}) or tetracycline (10 µg mL^{-1}) was added for the maintenance of plasmids, 1 mM isopropyl-β-D-thiogalactopyranosid (IPTG) was used to induce cells.

Table 1. Strains and plasmids used in this study

Strain or plasmid	Relevant feature	Reference/Source
Strains		
E. coli DH5α	supE44, ΔlacU169 (Ø80lacZΔM15), hsdR17, recA1, endA1, gyrA96, thi-1, relA1	Hanahan et al.[283]
E. coli W3110	prototroph	Bachmann et al.[284]
E. coli K27	fadD$^-$, deficient in fatty acid degradation	Overath et al.[240]
E. coli JMU194	fadR::Tn10, fadA30	Rhie et al.[285]
P. putida GPo1	Wild type, ATCC 29347	Schwartz et al.[286]
P. putida GPo500	PHA degradation mutant of GPo1	Huisman et al.[17]
Plasmids		
pGEM-T Easy	Cloning vector; lac operon, Amp^r	Promega (Madison, USA)
pJET1/blunt	Cloning vector; lac operon, Amp^r	Fermentas UAB (Vilnius, Lithuania)
pUCP26	Shuttle vector; broad-host range ori$_{1600}$, P$_{lac}$, Tc^r	West et al.[287]
pBTC2	phaC2 as 1.9 kb BamHI-KpnI fragment in pVLT35, P$_{lac}$, Sm/Sp^r	Ren et al.[190]
pGFPuv	gfp, P$_{lac}$, Amp^r	BD Biosciences Clontech (Palo Alto, USA)
pKR1	acs1 with SphI-and HindIII site as 1.7 kb fragment in pGEM-T Easy	This study
pKR2	acs1-gfp with SphI-and HindIII site as 2.4 kb fragment in pGEM-T Easy	This study
pKR3	acs1-gfp as 2.4 kb SphI-HindIII fragment in pUCP26	This study
pKR4	acs2 with SphI-and HindIII site as 1.7 kb fragment in pJET1/blunt	This study
pKR5	acs2-gfp with SphI-and HindIII site as 2.4 kb fragment in pJET1/blunt	This study
pKR6	acs2-gfp as 2.4 kb XbaI-HindIII fragment in pUCP26	This study
pKR7	gfp as 0.7 kb SphI-HindIII fragment in pGEM-T Easy	This study
pKR8	gfp as 0.7 kb SphI-HindIII fragment in pUCP26	This study
pKR9	gap between acs1 and acs2 in pJET1/blunt	This study

DNA manipulations

Isolation and purification of chromosomal and plasmid DNA, restriction digestion, ligation, gel electrophoresis, transformation and electroporation were carried out according to Sambrook and Russel.[288] Chromosomal DNA from *P. putida* GPo1 served as template for cloning of *acs*; pGFPuv served as template DNA for cloning *gfp*. Primers used in this study are listed in Table 6.2 and were provided by Microsynth (Microsynth AG, Balgach, Switzerland). For PCR, a total volume of 20 µL with the following components was used: 0.4 units Taq DNA polymerase, 0.2 mM dNTP, 0.2 mM forward primer, 0.2 mM reverse primer, 0.1-0.3 µg template DNA and 1xTaq DNA Polymerase reaction buffer (Bioconcept). PCR was performed on a thermocycler T3000 (Biometra, Goettingen, Germany) by 3 consecutive steps: Firstly, 3 min at 95°C, then 30 cycles of 30 sec at 95°C, 30 sec at a specific annealing temperature and an elongation at 72°C for a specific time, in the end 5 min at 72°C. DNA sequencing was carried out by Synergene GmbH (Schlieren, Switzerland).

Table 6.2. Primers used in this study

Primer	Sequence
w_1 forward	5'-TAGATGTGGTACAGCGGCAGCGG-3'
w_1 reverse	5'-CCGCTGCCGCTGTACCACATCTA-3'
w_2 forward	5'-GCTGCCAGTAGCCCTTCATCACCTGCGG-3'
w_2 reverse	5'-CCGCAGGTGATGAAGGGCTACTGGCAGC-3'
acs1-start	5'-CGCATGCATGATCGAAAATTTTTGGAAGG-3'
acs1-stop	5'-TAAGCTTTCAGGCGATCTTCTTCAA-3'
acs1-gfp-start	5'-CCCGCATGCATGATCGAAAATTTTTGGAAGGAT-3'
acs1-gfp-middle (forward)	5'-GCTTGAAGAAGATCGCCAGTAAAGGAGAAGAACTTTTCAC-3'
acs1-gfp-middle (reverse)	5'-GTGAAAAGTTCTTCTCCTTTACTGGCGATCTTCTTCAAGC-3'
acs-gfp-stop	5'-CCCAAGCTTTTATTTGTAGAGCTCATCCATG-3'
acs2-start	5'-ATAGCATGCATGCAAGCCGACTTCTGGAATGACA-3'
acs2-stop	5'-TTTAAGCTTTCACGCTATATCGCGCAACTCCCG-3'
acs2-gfp-middle (forward)	5'-GCGCGATATAGCGAGTAAAGGAGAAGAACTTTTC-3'
acs2-gfp-middle (reverse)	5'-TCTTCTCCTTTACTCGCTATATCGCGCAACTCCCG-3'
gfp-start	5'-CCCGCATGCATGAGTAAAGGAGAAGAACTTTTC-3'
gap1	5'-TCCGCGATTCGCTGCCGA-3'
gap2	5'-GTAGGCCGGCACCATCTTCT-3'

Construction of pKR1, pKR4, and pKR9

Having single A-overlaps due to the Taq DNA polymerase, the PCR product of *acs1* could be ligated into the single 3'-T overhangs of pGEM-T Easy, leading to pKR1 (Table 6.1). The PCR products of *acs2* and of the sequence between *acs1* and *acs2* amplified using primers gap1/gap2 were inserted into pJET1 by blunt-end ligation, respectively, resulting in pKR4 and pKR9 (Table 6.1). Nucleotide sequence data generated in this study were deposited in EMBL Nucleotide Sequence Database under accession number AM911678.

Construction of *acs-gfp* containing plasmids

The *acs1-gfp* fusion was obtained by a two-step PCR method. *Acs1* was amplified with the primers *acs1-gfp*-start/*acs1-gfp*-middle (reverse) and *gfp* with the primers *acs1-gfp*-middle (forward)/*acs-gfp*-stop in two separate reactions. The primer *acs1-gfp*-middle binds to both *gfp* and *acs*. The purified PCR products served as templates for amplification of the whole fusion gene with primers *acs1-gfp*-start/*acs-gfp*-stop. PCR fragments were purified and ligated into the single 3'-T overhangs of pGEM-T Easy, giving pKR2 (Table 6.1). The SphI-HindIII digested inserts were ligated into SphI-HindIII digested pUCP26, resulting in pKR3 (Table 6.1). As control during microscopy studies, *gfp* was obtained with the primers *gfp*-start/*acs-gfp*-stop, and incorporated into pUCP26, leading to pKR8 (Table 6.1) by applying the same strategy.

The *acs2-gfp* fusion was obtained by a similar approach. The gene *acs2* was amplified with the primers *acs2*-start/*acs2-gfp*-middle (reverse) and *gfp* with the primers *acs2-gfp*-middle (forward)/*acs-gfp*-stop in two separate reactions. The purified PCR products served as templates for amplification of the whole fusion gene with primers *acs2*-start/*acs-gfp*-stop. PCR fragments were purified and ligated to pJET1/blunt, generating pKR5 (Table 6.1). The XbaI-HindIII digested inserts were ligated into XbaI-HindIII digested pUCP26, resulting in pKR6 (Table 6.1).

PHA granule isolation and analysis of granule-associated proteins

PHA granules of *P. putida* GPo1 were isolated from the cells by density centrifugation as reported previously.[289] Samples of purified granules were mixed 1:1 (v/v) with SDS-PAGE (sodium dodecylsulfate polyacrylamide gel electrophoresis) loading buffer[288] and the bound proteins were separated on SDS-polyacrylamide gels as described previously.[290]

Acyl-CoA-synthetase activity assay

For testing the acyl-CoA-synthetase activity we used Ellmann's assays in which the depletion of CoA is monitored spectroscopically by the formation of yellow colored *p*-nitrothiophenol. The method was adopted from Kraak et al.[291] and carboxylic acids (C_4-C_{16}) or (*R*)-3-hydroxycarboxylic acids (C_4-C_{16}) were used as substrates. The assay reagents had the following end concentrations: 40 mM potassium phosphate, 1 mM $MgCl_2$, 1 mM ATP, 1 mM CoA, and 1 mM substrate. Conditions were set to pH 7.0 and 30°C. The assay was started by the addition of PHA granules. Aliquots of 90 µL were withdrawn at timed intervals, quenched with 20 µL of trichloro acetic acid and the concentration of CoA~SH determined by Ellmann's reagent (5',5'-dithiobis-(nitrobenzoic acid)). 1 mM of Ellmann's reagent was used to stain a 50 µL sample in a total volume of 1 mL at pH 7. Absorbance was measured at 412 nm (ε = 13.6 cm^2/µmol)). The molecular mass of the synthesized 3-hydroxycarboxyl-CoA was confirmed by liquid chromatography-mass spectroscopy (LC-MS) using positive spray ionization (Agilent 1100 series, Agilent Technologies, USA). The MS settings were as follows: Atmospheric pressure chemical ionization mode, positive ionization; fragmentor voltage, 50 V; gas temperature, 350°C; vaporizer temperature, 375°C; drying gas (N$_2$) flow rate, 4 L min^{-1}; nebulizer pressure, 0.023 N m^{-2}; capillary voltage, 2000 V; corona current, 6 µA.

Microscopy studies

Bacterial cultures were cultivated in 50 mL E2 medium supplement with 15 mM sodium octanoate. They were inoculated from precultures in TSB with a ratio of 1:50 and induced by 1 mM IPTG at the early exponential growth phase. For microscopy, aliquots were withdrawn in the late exponential or stationary growth phase. To immobilize bacteria cells on glass slides, 10 µL of bacterial suspension were mixed with 10 µL of 1% (w/v) agarose. Cells were stained with Nile red (Sigma-Aldrich, Buchs, Switzerland) as previously described to analyze the formation of PHA granules.[292] Microscopy study was performed on a Leica DFC350 FX (Leica Microsystems, Heerbrugg, Switzerland). Green fluorescence was excited at 450-490 nm and detected at 525-550 nm. Nile red fluorescence was excited at 515-560 nm and detected at 590 nm. Exposure time was 1.0-1.5 s for depicting GFP-fluorescence, and 0.5-1.0 s for red fluorescence. Bright field pictures were imaged for 10 ms.

Results and Discussion

Discovery of acyl-CoA synthetase activity on PHA granules

P. putida GPo1 accumulates PHA as granules during growth on fatty acids and under nitrogen limitation. These granules can be isolated by density centrifugation and used to study granule-associated proteins such as PHA polymerase and depolymerase.[291, 293] In this study, PHA granules were purified from GPo1 cells grown on octanoate and subsequently analyzed for activities of granule-associated proteins. Interestingly, an acyl-CoA synthetase (ACS1) activity was detected: When PHA granules were incubated with CoA, ATP, MgCl$_2$, Triton X-100, which are necessary for the activity of a typical ACS, and 3-hydroxynonanoic acid, a rapid depletion of CoA~SH could be observed. Concomitant, a product was formed, which was identified by LC-MS as 3-hydroxynonanoyl-CoA. When any of the reacting compounds such as CoA, ATP, MgCl$_2$, Triton X-100, or 3-hydroxynonanoic acid was omitted from the reaction mixture, or when the PHA granules were treated at 95°C for 5 minutes prior to the assay, there was almost no depletion of CoA~SH and no product was detected. This demonstrated that an enzymatic activity typical of an ACS was associated with PHA granules. The presence of an ACS on PHA-granules has not been reported so far despite the possibility that enzymes such as ACS are required to activate the monomeric acids released from the granules into their CoA-linked form. We speculated that the ACS1 identified here may serve this purpose.

Cloning of two acyl-CoA synthetase genes (*acs1* and *acs2*) of *P. putida* GPo1

To clone the ACS genes of *P. putida* GPo1, we compared sequences of *acs* from different *Pseudomonas* strains: *P. putida* U, *P. putida* F1 ctg307, *P. aeruginosa* PAO1, *P. aeruginosa* C3719, and *P. aeruginosa* PA7. Primers *acs1*-start and *acs1*-stop were designed and used in a PCR with GPo1 chromosomal DNA as template (Table 6.2, Fig. 6.1). This led to a 1.7 kb DNA product (*acs1*), whose sequence shares 94% identity to *fadD1* of *P. putida* U. From the Blast search, a second acyl-CoA synthetase upstream of *fadD1* is identified in all above mentioned *Pseudomonas* strains. To investigate whether there is also a second *acs* upstream *acs1* in *P. putida* GPo1, primers were designed based on the conserved regions in the *Pseudomonas* strains: primer w_1 forward and w_2 reverse (Table 6.2, Fig. 6.1).

```
                 start
ACS1   MIENFWKDKYPAGITAEINPDEFPNIQAVLKQSCQRFADKPAFSNLGKTITYGELYALSG      60
ACS2   MgadFWnDKrPAGvpstIdinaytsvveVferSCkRFADrPAFSNLGvTlTYaELerhSa      60
       m   fw dk pag    i             v    sc rfad pafsnlg t ty el     s

ACS1   AFAAWLQQHTDLKPGDRIAVQLPNVLQYPVAVFGAMRAGLIVVNTNPLYTAREMEHQFND     120
ACS2   AFAAWLQQHTDLKPGDRIAVQmPNVLQYPiAVFGAlRAGLIVVNTNPLYTeREMrHQFkD     120
       afaawlqqhtdlkpgdriavq pnvlqyp avfga raglivvntnplyt rem hqf d

ACS1   SGAKALVCLANMAHLAEKVVPKTQVRHVIVTEVADLLPPLKRLLINSVIKYVKKMVPAYN     180
ACS2   SGArALVyLnmfgkrvqeVlPdTgieylIearmgDmLPtaKgwLvNtVvvdkVKKMVPAYq     180
       sga alv l            v p t   i       d lp k   l n v  vkkmvpay

ACS1   LPRAVRFNDALALGKGQPVTEANPQANDVAVLQYTGGTTGVAKGAMLTHRNLVANMLQCR     240
ACS2   LPqAVsFkhvLrqGrelshkpvplsleDtAVLQYTGGTTGlAKGAMLThgNLVANMLQvl     240
       lp av f    l  g             d avlqytggttg akgamlth nlvanmlq

ACS1   ALMGSN......LHEGCEILITPLPLYHIYAFTFHCMAMMLIGNHNVLISNPRDLPAMV     293
ACS2   AcfsqhgpdgqklikdGqEvmIaPLPLYHIYAFTanCMcMMvtGNHNVLItNPRDisgfi     300
       a                ge i PLPLYHIYAFT cm mm  gnhnvli nprd
                                ←── W₁
ACS1   KELGKWKFSGFVGLNTLFVALCNNEAFRALDFSALKITLSGGMALQLSVAERWKAVTGCA     353
ACS2   KELGKWrFSallGLNTLFVALmdhpgFRqLDFSALKvTnSGGtALvkatAERWeAlTGCr     360
       kelgkw fs   glntlfval      fr ldfsalk t sgg al    aerw a tgc

ACS1   ICEGYGMTETSPVAAVNP.AEANQVGTIGIPVPSTLCKVIDDNGNELPLGEVGELCVKGP     412
ACS2   IvEGYGlTETSPVAstNPyggqlarlGTvGIPVagTafKVIDDdGNELPLGErGELCiKGP     420
       i egyg tetspva  np         gt gipv  t   kvidd gnelplge gelc kgp

ACS1   QVMKGYWQREEATAEILDSNGWLKTGDIAVIQPDGYMRIVDRKKDMILVSGFNVYPNELE     472
ACS2   QVMKGYWQqpEATAqaLDaeGWfKTGDIAVIdPDGftRIVDRKKDMIiVSGFNVYPNEiE     480
       qvmkgyw   eata  ld  gw ktgdiavi pdg   rivdrkkdmi vsgfnvypne e
        ──→ W₂
ACS1   DVLAALPGVLQCAAIGVPDEKSGEVIKVFIVVKPGMTVTKEQVMEHMRANVTGYKVPRQI     532
ACS2   DVvmghPkVanCAAIGVPDErSGEavKlFvVpreGg.lsvdelkayckANfTGYKVPkhI     539
       dv     p v  caaigvpde sge   k f v  g            an tgykvp   I
                                                stop
ACS1   EFRDALPTTNVGKILRRELRDEELKKQGLKKIA                                565
ACS2   vlResLPmTpVGKILRRELRDla..........                                562
       r   lp t vgkilrrelrd
```

Figure 6.1 Amino acid alignment for CoA synthetases ACS1 and ACS2 (accession no. AM911678) from *P. putida* GPo1. Conserved regions that provided the basis to derive primers w_1 (forward and reverse) and w_2 (forward and reverse) for performing genomic walking are underlined. Sequences for primers to clone the entire *acs1* and *acs2* are indicated in light grey, labeled as start and stop.

PCR with these primers resulted in a PCR product of approximately 740 bp length and the nucleotide sequence of this product was determined. Starting from this fragment, the upstream and downstream regions were analyzed with the primers w_1 reverse and w_2 forward (Table 6.2, Fig. 6.1) and with the help of a GenomeWalker Universal Kit (Clontech, CA, USA), and thus the sequence of a whole gene (*acs2*) was obtained. The sequence shares 92% identity with *fadD2* from *P. putida* U.[248] Primers *acs2*-start and *acs2*-stop (Table 6.2) were further designed to amplify the entire *acs2* gene of GPo1 by PCR and the product was sequenced to confirm the obtained *acs2* sequence.

Primary structures were calculated with DNAMAN (Lynnon Corporation, Quebec, Canada). The *acs1* encoded polypeptide has 565 amino acids with a predicted molecular mass of 61.78 kD. The deduced amino acid sequence of ACS1 of *P. putida* GPo1 exhibited a high identity (98%) to that of *P. putida* U, and had significant homology to a long chain fatty acid CoA thiolase FadD of *E. coli* K12 (54% identity). The primary structure of ACS1 contained a putative ATP-binding region (T215GGTTGVAKG),[294] a putative CoA-binding site (V461SG)[245] and a putative signature motif (N432GWLKTGDI....IVDRKK), which modulates fatty acid substrate specificity.[239] The *acs2* encoded polypeptide has 562 amino acids with a predicted molecular mass of 61.66 kD. Its deduced amino acid sequence showed 61% identity with ACS1, 94% identity with its homologue in *P. putida* U, and 53% identity with FadD of *E. coli*. In ACS2, we also detected a putative ATP-binding region (T215GGTTGLAKG), a putative CoA-binding site (V469SG), and a signature motif (E440GWFKTGDI...IVDRKK). From calculation of the primary structures ACS1 is slightly more hydrophobic compared with ACS2, and has three predicted transmembrane segments (amino acids 79-107, 253-281, 298-324), out of which only two were found in ACS2 (amino acids 79-107, 305-331). At present, it is not clear why *Pseudomonas* strains have two acyl-CoA synthetases, while *E. coli* contains only one which is involved in fatty acid degradation. It is possible that the presence of multiple homologues of an enzyme may permit them to perform non-overlapping functions. It might also be possible that the homologues of an enzyme may play the auxiliary role: when the enzyme is lacking, its homologues can substitute its function, as such described for FadD1 and Fad2 in *P. putida* U.[248]

Complementation of a *fadD*-defective *E. coli* mutant

When long-chain fatty acids serve as sole energy and carbon source, growth of *E. coli* requires FadD to activate these compounds. Fatty acids are used for both the β-oxidation and the synthesis of membrane phospholipids.[240] To determine whether the ACSs from *P. putida* GPo1 are sufficient for growth of *E. coli* on fatty acids, the *fadD*-defective *E. coli* K27 was equipped with pRK3 containing *acs1-gfp* or with pKR5 containing *acs2-gfp* (Table 6.1). The recombinants were streaked on medium E2 agar plates supplemented with 2 mM octadecanoic acid (C18) dissolved in 0.5% Brij56. *E. coli* W3110[284, 295] with intact β-oxidation enzymes was used as positive control. *E. coli* K27 equipped with the empty vector was used as negative control. For comparison, *E. coli* cells were also cultivated on plates containing 25 mM glucose. As expected, all strains were able to grow on glucose. Growth of *E. coli* W3110 on C18 plates was observed after 2-3 days. *E. coli* K27 transformed with the empty vector was unable to grow on C18, even after prolonged incubation of 6 days. K27 harbouring *acs1* or *acs2* grew on C18 after incubation of 3-4 days. This demonstrated that both ACS1 and ACS2 from *P. putida* GPo1 could complement the function of FadD in *E. coli*. These results are consistent with the previous proposition that the two homologues (FadD1 and FadD2) of ACS1 and ACS2 in *P. putida* U are involved in degradation of fatty acids,[248] and both ACSs of GPo1 might supplement each other in converting fatty acids.

Subcellular localization of ACS1 and ACS2 of *P. putida* GPo1

In order to verify the cellular localization of ACSs *in vivo* in *P. putida* GPo1, a C-terminal fusion of ACS with the GFP was constructed and cloned into the broad-host-range vector pUCP26 as described in Materials and Methods. As control, pUCP26 carrying only *gfp* gene was used. The recombinants were cultivated in minimal medium with octanoate as the sole carbon source. Cells were investigated for PHA granule formation and GFP expression with light and fluorescence microscopy.

PHA granules were visible in GPo1 equipped with either GFP (Fig. 6.2A1), ACS1-GFP (Fig. 6.2A2) or ACS2-GFP (Fig. 6.2A3) after cells reached late exponential growth phase. In GPo1(*gfp*) cells, the green fluorescence of GFP was evenly distributed within the cytoplasm (Fig. 6.2B1). In GPo1(*acs2-gfp*), green fluorescence was mainly located along the cell membrane (Fig. 6.2B3), indicating that ACS2 is a cell membrane associated protein, like FadD in *E. coli*.[233] Interestingly, in GPo1(*acs1-gfp*) green fluorescence mainly occurred on the inclusion bodies (Fig. 6.2B2). These inclusion bodies could not be ACS1-GFP protein inclusions for the following reasons: Firstly, expression of ACS1-GFP in GPo1 was analyzed by SDS-PAGE and no detectable overproduction band was obtained compared to GPo1 cells, indicating that no protein inclusion bodies were formed; secondly, they also existed in GPo1(*gfp*) and GPo1(*acs2-gfp*)cells (Fig. 6.2A1 and 6.2A3); thirdly, they could be stained by Nile red, which is a typical feature of PHA and of other lipids: red fluorescent granules were observed in the cells carrying plasmids with *gfp* (Fig. 6.2E1), *acs1-gfp* (Fig. 6.2E2), or *acs2-gfp* (Fig. 6.2E3). The red fluorescent inclusion bodies concomitantly exhibited the green fluorescence (Fig. 6.2B2). These results strongly suggest that there is co-localization of ACS1 with PHA granules. Further incubation of the recombinants until stationary growth phase resulted in elongation of the cells (Fig. 6.2C1-3). In this case, the co-occurrence of PHA granules and the green fluorescence of ACS1-GFP fusion protein became even more distinct (Fig. 6.2C2 and 6.2D2), whereas the cells containing only GFP or ACS2-GFP exhibited the green fluorescence homogenously within the cells (Fig. 6.2D1) or along the cell membrane (Fig. 6.2D3), although the PHA granules were clearly visible by light microscope (Fig. 6.2C1 and 6.2C3). Cells were induced at early exponential phase, thus PHA had already been accumulated and ACS1-GFP fusion protein was mainly located at the PHA granules. In order to test its location in absence of PHA granules, cells were grown on citrate where no PHA formation is possible in GPo1. In these tests, ACS1-GFP fusions were detected in the cytosol and on the cell membrane (data not shown).

Previously, it has been reported that *in vitro* localization based on separated cell fractions can be misleading, because during cell disruption process membrane or granule-associated proteins can be detached and released into soluble cell fraction, and/or membrane fractions can be contaminated by non-membrane or no-granule-associated proteins.[296] Therefore, to determine the true localization of ACS1 and ACS2 *in vivo*, C-terminal fusions of ACS with GFP were constructed. The resulting fusion proteins were expressed under *in vivo* conditions and the fluorescent microscopy data demonstrated that the ACS1 is predominantly located on PHA granules, whereas ACS2 is on the cell membrane.

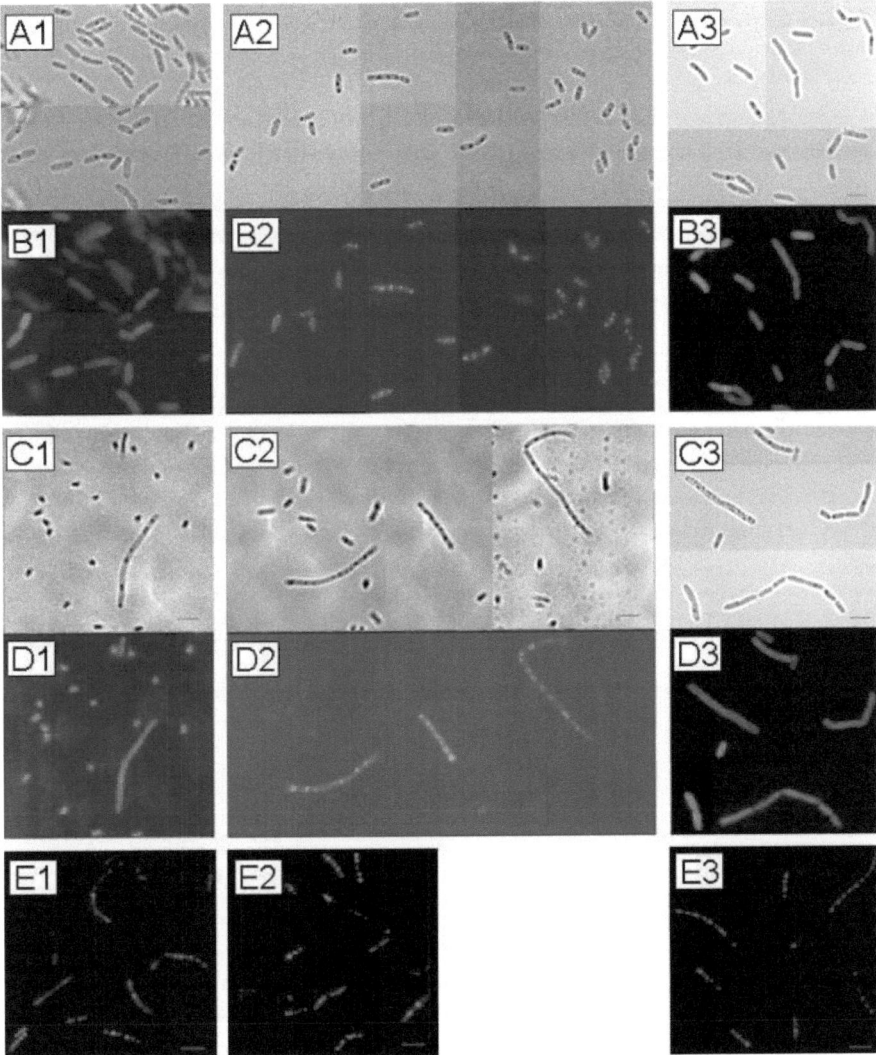

Figure 6.2 Phase-contrast (A and C) and fluorescence (B, D and E) microscopic analysis of recombinant *P. putida* GPo1 cells harbouring *gfp* (1), *acs1-gfp* (2) or *acs2-gfp* (3). The presence of PHA granules in GPo1(*gfp*) (A1 and C1), GPo1(*acs1-gfp*) (A2 and C2) and GPo1(*acs2-gfp*) (A3 and C3) were revealed with a bright field filter. With a GFP-specific filter, GPo1(*gfp*) cells showed evenly distributed green fluorescence (B1 and D1), whereas fluorescence mainly occurred at the inclusions in GPo1(*acs1-gfp*) (B2 and D2). In GPo1(*acs2-gfp*) green fluorescence was mainly distributed along the cytoplasmic membrane (B3 and D3). PHA granule formation in GPo1(*gfp*) (E1), GPo1(*acs1-gfp*) (E2) and GPo1(*acs2-gfp*) (E3) was confirmed by Nile Red staining. The bar represents 3.5 µm.

We further confirmed the localization of ACS1 by using *E. coli* recombinant. Wild-type *E. coli* is not able to accumulate PHA, and its genome contains no PHA polymerase or depolymerase, but an acyl-CoA synthetase (FadD). Previously it has been reported that pBTC2 containing PHA polymerase 2 of *P. putida* GPo1 enabled *E. coli* JMU194, a fatty acid degradation defective mutant, to synthesize PHA.[191] If the localization of ACS1 on PHA granules is an intrinsic characteristic of the ACS1, the FadD of *E. coli* should also locate on the PHA granules produced in *E. coli*, besides other sub-locations. Alternatively, if the ACS1 of GPo1 is specific for PHA metabolism and is absent in *E. coli*, the FadD can not complement the activity of ACS1 and thus is not located on the PHA granules.

In this study, mcl-PHA were produced in *E. coli* JMU194 (pBTC2) grown on hexadecanoate. The PHA granules were subsequently isolated and assayed for acyl-CoA synthetase activities. Neither any medium- nor long-chain length acyl-CoA synthetase activity could be detected, suggesting that the presence of this enzyme is specific for *Pseudomonas*.

Substrate specificity of ACS1 in *P. putida* GPo1

The substrate specificity of PHA granule-associated ACS1 was investigated (Table 6.3). *P. putida* GPo500 was used here due to the lack of a functional PHA depolymerase,[17] thus avoiding the interference of the PHA degradation products on ACS1 activities. PHA granules were isolated from GPo500 grown on octanoate. 3-hydroxycarboxylic acids and aliphatic fatty acids with different chain lengths were used as substrates. The highest activity obtained in this study was set as 100%.

Table 6.3. Substrate specificity of the PHA granule-associated acyl-CoA synthetase (ACS1). Relative activities are shown (%) and the highest activity was set to 100%. C_4-C_{16} symbolize carboxylic acids with a chain length of 4 to 16 carbon atoms; 3-OH-(C_4-C_{16}) symbolize the corresponding 3-hyroxycarboxylic acids.

substrate	activity	substrate	activity
C_4	1	3-OH-C_4	2
C_5	2	3-OH-C_5	5
C_6	11	3-OH-C_6	12
C_8	62	3-OH-C_8	47
C_{10}	74	3-OH-C_{10}	51
C_{12}	100	3-OH-C_{12}	53
C_{14}	86	3-OH-C_{14}	58
C_{16}	74	3-OH-C_{16}	66

Our results showed that ACS1 of GPo1 had a broad substrate specificity with high affinity for medium- and long-chain 3-hydroxycarboxylic acids (C8-18) (Table 6.3). Low activities were observed with 3-hydroxybutyric acid, 3-hydroxyvaleric acid and 3-hydroxyhexanoic acid. It was found that ACS1 was not specific only for 3-hydroxycarboxylic acids; it had even higher activities towards non-substituted fatty acids (Table 6.3). Maximum activities were observed for dodecanoic acid, after which the activity decreased with longer chain length alkanoates. This decrease is probably not related to a decrease in affinity but due to the insolubility of tetradecanoic acid and hexadecanoic acid. Thus, ACS1 might be functional in two processes *in vivo*. Firstly, it activates fatty acids when they enter the cell to make them accessible for β-oxidation.[233] Secondly, it also activates hydroxycarboxylic acids that are released from PHA granules by PHA depolymerase during PHA degradation. It is probable that ACS1 of GPo1 is only active when located at a membrane (either at the cellular membrane or at the phospholipid monolayer which engulfs PHA granules) as it is the case for FadD in *E. coli*[233] and many other FadD homologues.[297]

Modelling of the PHA metabolism

Taking advantage of the knowledge acquired previously and our findings in this study[167, 264], we developed a model to gain insight into the metabolism of PHA in *P. putida* GPo1 (Fig. 6.3). The model shows that when cells facing carbon excess and nutrient limitation they store the extra carbon in the form of PHA via PHA polymerase.[298] Under carbon starvation conditions, PHA depolymerase degrades PHA and releases 3-hydroxycarboxylic acid monomers.[46, 264, 270] The released monomers are then activated to hydroxyacyl-CoAs by ACS1 via an ATP-dependent reaction.

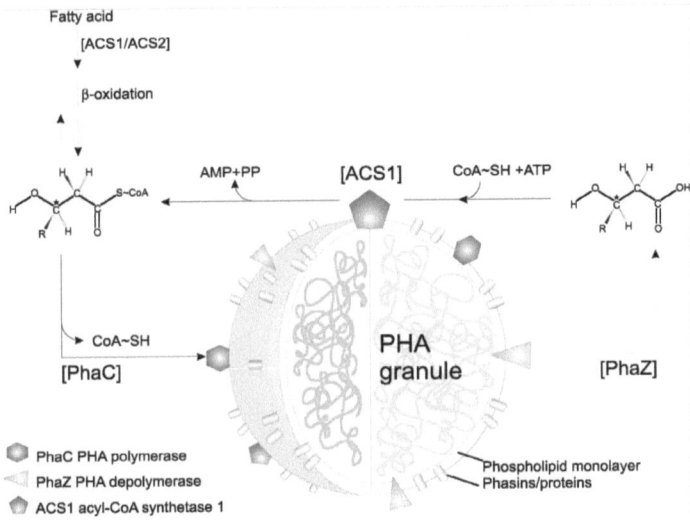

Figure 6.3 Model for PHA metabolism at the enzymatic level.

The metabolite is a substrate for PHA polymerase as well as the fatty acid β-oxidation cycle: depending on the metabolic state of the cell, the hydroxyacyl-CoAs will either be incorporated into nascent PHA polymer chains by the PHA polymerase, or will be oxidized by the β-oxidation pathway. PHA synthesis and degradation have been shown to be a simultaneous process,[40, 41] which seems to be a futile cycle of PHA turnover. The metabolic advantage of such a mechanism is doubtful as a significant amount of energy is wasted. One explanation is that PHA is a buffer of reducing power and carbon. Depending on the nutrition and the cellular demand for carbon and reducing power, the net balance between PHA synthesis and degradation is positive (PHA accumulation) and negative (PHA degradation). For efficient turnover of PHA, it is plausible that the critical enzymes involved in the cyclic metabolic pathway of PHA, namely PHA polymerase, depolymerase and acyl-CoA synthetase, are associated with PHA granules as illustrated in Figure 6.3.

Summary

In contrast to the efforts made for understanding the enzymatic hydrolysis of short-chain-length PHA,[15, 196-198, 299] the biodegradation of medium-chain-length PHA has been rarely studied.[15] Recently the mcl-PHA depolymerase has been purified and characterized at the biochemical level.[46] The role of the depolymerase has been demonstrated both *in vivo* and *in vitro* to hydrolyze PHA to (*R*)-hydroxycarboxylic acid monomers;[46, 264] however, how the monomers are utilized by cells has not been studied. It was postulated that the monomers are first activated by an acyl-CoA synthetase to a CoA-linked form which is either an intermediate of β–oxidation or incorporated back into PHA.[40, 61, 300] In this report, we demonstrated for the first time the subcellular localization of an acyl-CoA synthetase on PHA granules through both *in vitro* and *in vivo* experiments. At present, it is not clear why two ACSs exist in *P. putida*. The construction of mutants in which one of the two *acs* genes is lacking will facilitate to understand the function of the ACSs. Furthermore, to gain more insight on the properties of the acyl-CoA synthetases identified in this study, purification of these enzymes is necessary. The purified enzymes will allow the examination of the activities and substrate specificities on a quantitative level.

Acknowledgement

We wish to thank Dr. Eva Brombacher and Birsen Demirbag for practical help with molecular cloning experiments. Thanks are given to Dr. Manfred Zinn and Dr. Linda Thöny-Meyer for reading the manuscript.

Chapter 7

Investigation of an acyl-CoA synthetase knockout mutant of *Pseudomonas putida* GPo1

Abstract

In order to investigate pathways of medium-chain-length poly(hydroxyalkanoates) (mcl-PHA) metabolism in *Pseudomonas putida* GPo1 acyl-coenzyme A synthetase 1 (ACS1) was knocked out. The phenotype of the mutant was studied with respect to growth and PHA synthesis and degradation: When grown on citrate, the mutant exhibited a phenotype similar to the wild type; when fatty acids were used as the sole carbon source growth of the mutant was impaired and only observed after a prolonged lag-phase of 12 hours. Subsequent growth rates and cell-dry-weights reached were similar to the wild type. Furthermore, the ACS1 mutant accumulated 30-50% less PHA than its parental strain GPo1. Under carbon starvation conditions PHA degradation was faster in the mutant: After 6 hours only ~35% of initial PHA were found in the knockout mutant, whereas more than ~50% of initial PHA were left in the wild type. It seems that ACS1 plays an important role in PHA metabolism.

Introduction

Biodegradable and biocompatible polyesters such as poly(hydroxyalkanoates) (PHA) gained considerable attention during last decades. Many bacteria accumulate PHA when a carbon source is in excess and growth is limited by the lack of other nutrients.[13, 21, 166, 174] When the cells are exposed to carbon limitation, the accumulated PHA can be degraded to monomers, which can be re-utilized by the bacteria as a carbon and energy source.[21] *Pseudomonas putida* GPo1 is a medium-chain-length (mcl) PHA producer incorporating a wide variety of monomer units, mainly depending on the available carbon source, e.g., fatty acids. When *P. putida* GPo1 is grown on octanoic acid, PHA are composed of *R*-3-hydroxyoctanoate and *R*-3-hydroxyhexanoate units (Fig. 7.1).[301]

In *P. putida*, PHA are built up by two polymerases PhaC1 and PhaC2 with similar substrate specificity.[302] Degradation of PHA is enhanced under carbon starvation and is catalyzed by PHA depolymerase (PhaZ). PhaZ of *P. putida* KT2442 is one of the best studied PHA depolymerases up to now.[46] PhaZ degrades PHA into monomer acids.[46] Released PHA monomers are assumed to be activated by CoA to enter the β-oxidation as well as to be channeled back to PHA synthesis.[40, 271] The cyclic nature of PHA metabolism, hence, the simultaneous synthesis and breakdown of PHA has been shown, e.g., by radioactive labeling experiments.[40, 41]

In order to gain insight into PHA metabolism, fruitful approaches investigated reaction pathways of the common precursor 3-hydroxyacyl-CoA that interconnects β-oxidation with PHA metabolism.[303, 304] Experiments often involve blocking the formation or degradation of 3-hydroxyacyl-CoA, by knocking out the corresponding genes and studying the resulting phenotypes. For example, deletion of 3-ketoacyl-coenzyme A (CoA) thiolase (*fadA*) and 3-hydroxyacyl-CoA dehydrogenase (*fadB*) in *P. putida* KT2442 weakened the β-oxidation pathway.[304] PHA was produced with the major monomer component being the hydroxycarboxylic acid structurally related to the fatty acid feed. Liu et al. estimated that the complete removal of the β-oxidation activity could lead to the formation of homopolymers consisting only of the fatty acids related structures.[304] The *fadBA* operon in *P. putida* U encodes an enzymatic system (including 3-hydroxyacyl-CoA dehydrogenase and 3-ketoacyl-CoA thiolase) that is responsible for the β-oxidation pathway of aromatic fatty acids.[248] Deletion of fadBA resulted in an over-producer strain of poly-3-hydroxy-n-phenylalkanoate.[224, 248]

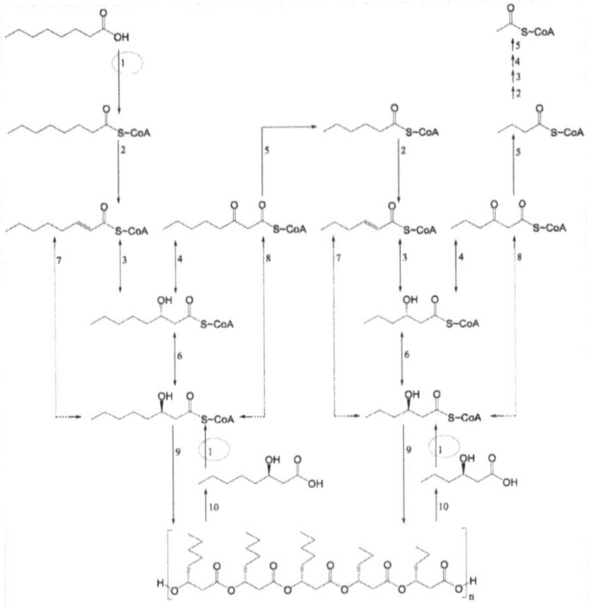

Figure 7.1 Metabolic pathways of P. putida showing synthesis of poly(R-3-hydroxyoctanoate-co-R-3-hydroxyhexanoate) when grown on octanoic acid. Hypothetical reactions are indicated with dashed arrows. Reaction conditions are the following (adapted from Kunau et al.[180]) 1: acyl-CoA synthetase 1 or 2 (+HS~CoA, +ATP, -AMP/PP, -H₂0), encircled; 2: acyl-CoA dehydrogenase (+FAD, -FADH₂); 3: enoyl-CoA hydratase (+H₂0); 4: NAD dependent S-3-hydroxyacyl-CoA dehydrogenase (+NAD, -NADH₂); 5: 3-ketothiolase (+HS~CoA, -acetyl-S~CoA); 6: 3-hydroxyacyl-CoA epimerase; 7: R-specific enoyl-CoA hydratase (-H₂0); 8: NADPH dependent 3-ketoacyl-CoA reductase (+NADP, -NADPH₂); 9: PHA polymerase (-HS~CoA); 10: PHA depolymerase (+H₂0).

In P. putida KCTC1639, fadA (3-ketoacyl-CoA thiolase) was blocked to induce a metabolic flux of the intermediates generated from the β-oxidation pathway for the direct biosynthesis of mcl-PHA, because degradation of the carbon chains of these intermediates cannot proceed further.[305] The fadA knockout showed increased PHA accumulation when succinate was added to stimulate cell growth.[305]

One approach to modify the concentration of PHA precursor 3-hydroxyacyl-CoA is to manipulate fadD, encoding acyl-CoA synthetase. These enzymes are responsible for activation of free carboxylic acids (e.g., PHA monomers) to enable their further conversion. So far, two fadD genes coding for CoA synthetases have been discovered in P. putida U. When fadD1 was knocked out growth on fatty acids was delayed, leading to the assumption that fadD2 might become induced when fadD1 is inactivated.[248] PHA accumulation in the fadD1-negative mutant was similar to its parental strain.[248] The

Chapter 7

construct of the corresponding double knockout of *fadD1* and *fadD2* could not be obtained and it was concluded that this mutant might not be viable.[248] It has not yet been possible to keep apart the functions of these two gene products.

In this study, we knocked out one (*acs1*) of the two *acs* identified in *P. putida* GPo1. ACS1 is assumed to ligate CoA to the carboxylic groups of PHA monomers in order to activate them for catabolic processes such as β-oxidation. The *acs1* gene was disrupted by means of insertion of a kanamycine resistance gene. The phenotype was studied regarding growth behavior and the ability to accumulate and degrade PHA. Except a prolonged lag-phase, growth characteristics of the mutant were similar to the parental strain; however the PHA content was significantly reduced and PHA degradation was accelerated in the mutant. The knowledge obtained in this study enabled us to gain insight into metabolic pathways of PHA synthesis and degradation.

Materials and methods

Bacterial strains

Pseudomonas putida GPo1 was used as model organism for mcl-PHA producers.[178] *Escherichia coli* DH5α[283] was used as host for plasmid construction and *E. coli* S17-1 as a vector donor in conjugation (contains *tra* genes of plasmid RP4 in the chromosome; *recA*; *thi*-1).[306]

Plasmid construction

The gene *acs* was amplified from genomic DNA of *P. putida* GPo1 using the following primers: *acs1*-start 5'-CGCATGCATGATCGAAAATTTTTGGAAGG-3' and *acs1*-stop 5'-TAAGCTTTCAGGCGATCTTCTTCAA-3' and applying a standard pcr with 30 cycles of melting (95°C/30s), annealing (55°C/30s), and elongation (72°C/60s). The obtained 1.7 kb fragment was cloned into a pGEM-T Easy vector (Promega, Madison, USA).

A kanamycine resistance gene and its promoter (km) were obtained from pUTmini-Tn5 Km (Biomedal, Sevilla, Spain) by standard pcr with 64°C annealing temperature and with the following primers: km-start 5'-GGGCCATGGGGTAAGGTTGGGAA-3' and 5'-GGGCCATGGTCAGAAGAACTCGTCAA-3'. NcoI restriction sites (introduced at the beginning and the end of the fragment by the primers) were used to insert *km* into NcoI restricted *acs* (Fig. 7.2). The resulting deletion *acs::km* was cloned into pUC18Not and subsequently into NotI restricted suicide vector pUTmini-Tn5 Tc (containing a tetracycline resistance gene (tc)) by the means of the PANC-1plus mini-Tn5 vector kit (Biomedal) (Fig. 7.2).

Double homologous recombination for construction of *acs* knockout

Plasmid pUTmini-Tn5 Tc *acs::km* was transformed into *E. coli* S17-1.[288] For biparental mating, *E. coli* S17-1 harboring the suicide plasmid and wild type *P. putida* GPo1 (both suspended in 0.9% NaCl solution to OD_{600} = 4.0) were mixed at a ratio of 1:3 and incubated on a rich media agar plate at 30°C for 3 hours. Afterwards, the material was diluted and plated onto selective plates. Single colonies exhibiting tc and km resistance were isolated and identified as single crossover mutants containing the whole suicide vector (Fig. 7.2).

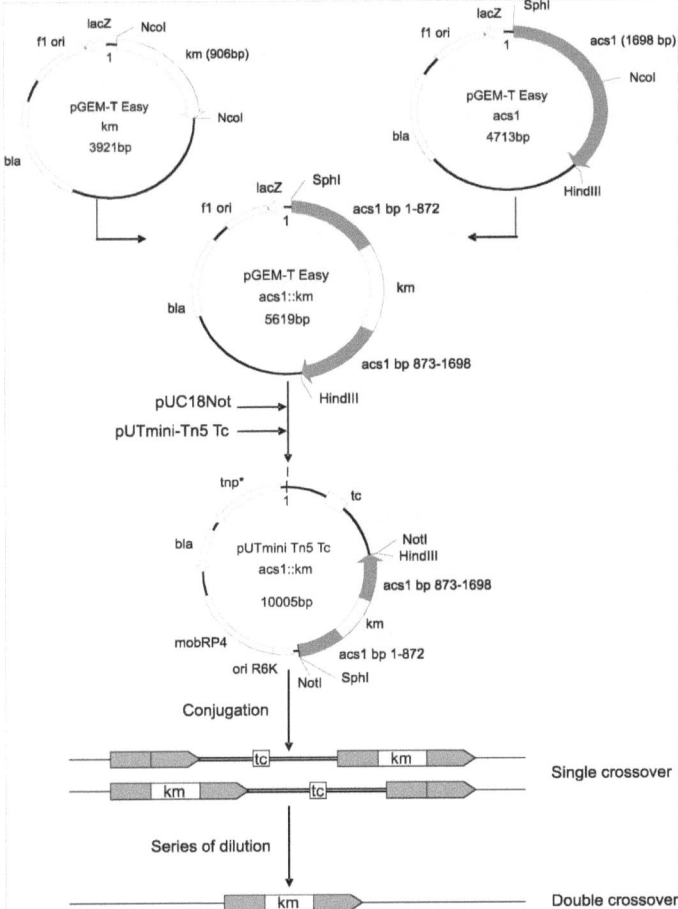

Figure 7.2 Schematic representation for construction of an *acs* knockout mutant *P. putida* GPo1. The following genes are indicated: acs (acyl-CoA synthetase 1 from *P. putida* GPo1), *bla* (ampicillin resistance), f1 ori (phage f1 region), *lacZ* (*lacZ* start codon), *tnp** (transposase), *mob*RP4 (origin of transfer), ori R6K (origin of replication), *km* (kanamycine resistance), *tc* (tetracycline resistance). The suicide vector was transferred to *P. putida* GPo1 by conjugation. Series of dilution served to select for double crossover events.

In order to obtain double crossover mutants, series of dilution were carried out (Fig. 7.2): Single crossover mutants were grown overnight on rich media without antibiotics. After 24 hours, they were diluted 1:100 to fresh media. This process had to be repeated 7 times to observe a spontaneous looping out of the remaining vector backbone. Colonies exhibiting tc sensitivity and km resistance were identified as double crossover mutants. Mutants were analyzed for fluorescence on King's agar (indicating presence of *P. putida*). The genotype was analyzed by colony pcr with previously mentioned primers.

Cultivation conditions

Wild type and mutants were grown in batch cultures on mineral media E2[62] containing either 10 mM sodium citrate or 15mM sodium octanoate as carbon source. Cultures were inoculated 1:100 from overnight pre-cultures in tryptic soy broth and were cultivated in 50-70 mL media with 110 rpm at 30°C.

Determination of PHA content and composition

The method was adapted from protocols for the quantification of fatty acids.[255, 307] Freeze-dried cell pellets (5-50mg) were transferred into 10 mL Pyrex tubes. An internal standard (10 gL^{-1} of 2-ethyl-2-hydroxybutyric acid in 1 mL of methylene chloride) was added. After dissolving the sample at room temperature for 1 hour, 1 mL of boron trifluoride (BF_3) in methanol (1.3 M, puriss., Fluka AG, Buchs, Switzerland) was added for conversion. The tube was tightly sealed and the mixture was vigorously shaken. Subsequently, samples were heated to 80°C for 16-20 hours. After cooling to room temperature, the reaction mixture was extracted twice with 2 mL of saturated NaCl. The organic phase was dried over Na_2SO_4, neutralized by adding Na_2CO_3, and filtered through a 1.0 µM nylon filter. Samples were derivatized twice for independent analysis.

Samples (1µL) were analyzed on a GC (GC 8575 Mega 2, Fisons Instruments, Rodano, Italy) equipped with a Supelco SPB-35 (30m×0.32 mm, film thickness 0.25µm) column (Supelco, Bellefonte, USA) and a flame ionization detector. Samples were injected with a split ratio of 1:10 at an initial temperature of 80 °C. The temperature was increased with a rate of 10°C min^{-1} to 240°C. For calibration, known amounts of pure 3-hydroxybutyric, -valeric, -hexanoic, -octanoic, -nonanoic and -undecanoic acid (purchased from Larodan, Malmö, Sweden) were derivatized and measured to calculate their response factors. Standard deviations were 5-10%.

Results and discussion

Generation of *acs1* knockout mutant of *P. putida* GPo1

The *acs* gene, involved in PHA metabolism, was blocked by homologous recombination based on inserting a kanamycine resistance cassette into the gene in the chromosomal DNA of *P. putida* GPo1. Mutants were obtained by conjugation of *P. putida* GPo1 with *E. coli* S17-1 containing plasmid pUT-mini-Tn5 Tc *acs::km* and subsequent series of dilution (Fig. 7.2). After conjugation, we observed that the double-crossover event is extremely rare (no positive clones in 1000 colonies), however 25% of all clones incorporated the plasmid into their genome by single-crossover. Proceeding with single crossover mutants, ~12% became double crossover mutants after 7 series of dilution. Isolated double crossover mutants were growing on kanamycine, sensitive to tetracycline (the resistance of the vector backbone), and fluorescent on King's agar (indicating presence of *P. putida* and not a contaminating bacteria strain), and no plasmids could be isolated from cells (data not shown). Hence, phenotype confirmed the event of a double crossover. The genotypes of double homologous recombinants were confirmed by colony PCR with previously mentioned primers *acs1*-start and stop (Fig. 7.3).

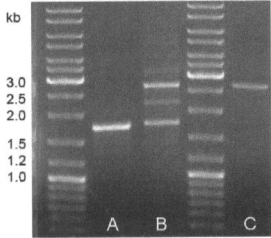

Figure 7.3 Colony pcr with primers on start and stop of *acs* gene. The lengths of markers are indicated in kb (kilo base pairs). A: Wild type having one *acs* gene (1.7 kb); B: Single crossover mutant harboring one deleted (2.7 kb) and one wild type copy of *acs* (1.7 kb); C: Double crossover mutant having only one deleted *acs* gene (2.7 kb).

Growth behavior of the *acs1* knockout mutant

The resulting *acs* knockout mutant was cultivated in mineral medium E2 supplemented with citrate as carbon source. Growth characteristics and maximally reached optical densities did not differ with respect to the parental strain GPo1 (Fig. 7.4A). Utilization of citrate does not need the enzymes involved in fatty acid degradation, such as acyl-CoA synthetases, thus similar growth behavior of knockout and wild type were expected.

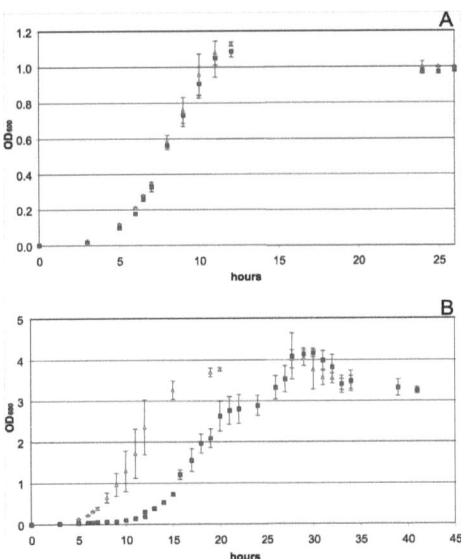

Figure 7.4 Growth on mineral medium with citrate (A) or octanoate (B) as carbon source (wild type: triangle; knockout: squares). Data represent the average of three independent measurements, and error bars denote standard deviation.

However, when grown on octanoate as the sole carbon source, knockout mutant showed a significantly longer lag-phase of ~12 hours (Fig. 7.4B) compared to the parental strain (~5 hours). This might indicate that fatty acid uptake or assimilation might be influenced in the knockout mutant and some alternative pathways have to be activated first. Once adapted after the initial lag phase, the mutant showed similar growth rates and final cell densities as to the wild type. Hence, we assumed that uptake and activation of fatty acids as carbon and energy source and β-oxidation pathways are not impaired in the mutant after overcoming the initial lag phase.

PHA accumulation and degradation of the *acs1* knockout mutant

The PHA content was analyzed during different growth phases. The main finding was that ACS knockout mutants accumulated less PHA than the wild type. When grown on octanoate in a batch culture, the patterns of PHA accumulation and degradation are basically the same in both strains, with increasing PHA content ensuing cell growth after depletion of nitrogen, followed by a slow decrease of PHA concentration in stationary cells (Fig. 7.5). However, total amount of PHA was 30-50% lower in the mutants (Fig. 7.5). The ratio of 3-OH-hexanoate (C6) and 3-OH-octanoate was analyzed to get hints why less PHA was accumulated. A shift in this ratio might indicate changes at β-oxidation processes, whereas a similar ratio could signify a hampered uptake of fatty acids. Here, we observed an overall C6:C8 ratio of 8.28 ± 1.6 (w/w) for the mutant and 7.44 ± 1.0 (w/w) for the wild type. However, it remains questionable whether this shift is significant or within the error range of experiments.

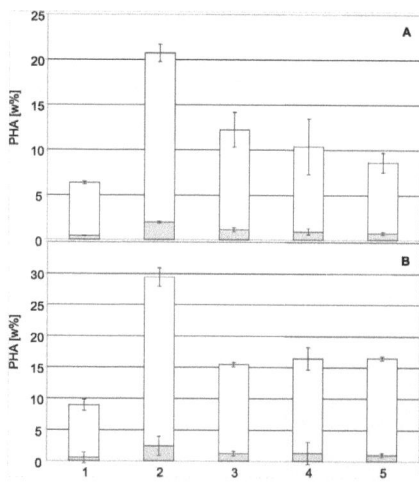

Figure 7.5 Effect of ACS knockout on PHA content. Strains were cultivated in mineral medium with 15 mM octanoate (A: knockout; B: wild type). PHA content is displayed in weight% (w%) of total cell dry mass, composed of the monomers octanoate (white) hexanoate (grey). The following growth phases were analyzed: 1. Early exponential (after 10h for wild type (wt), after 16h for knockout (ko)); 2. Late exponential (18h wt, 24h ko), 3. Start of degradation (27h wt, 32h ko), and late degradation phases 4. (40h) and 5. (52h).

In order to elucidate PHA degradation abilities in more detail, the mutant and the wild type strain were grown on octanoate (as above) till late exponential phase, then cells were harvested by centrifugation and the pellets suspended in the same volume of mineral medium without carbon source to force PHA degradation. PHA content preceding exposure to degradation conditions (i.e., initial PHA content) was 10.56 ± 1.65 w% in the mutant and 20.21 ± 12.43 w% in the wild type. Starting with similar amounts of cells (cell dry weight = 0.72 ± 0.28 gL^{-1}; OD$_{600}$ = 1.9 ± 0.3) both strains were able to degrade PHA, and the speed of PHA degradation was faster in the mutant (Fig. 7.6). Six hours after triggering PHA degradation, only ~35% of initial PHA were found in the knockout mutant whereas more than ~50% of initial PHA were left in the wild type. However, both strains did not completely decompose PHA, with ~20% of initial PHA remaining after 24 hours.

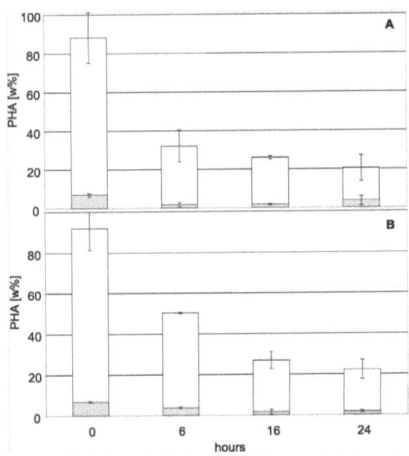

Figure 7.6 Degradation of PHA in minimal medium (A: knockout; B: wild type). At time zero, PHA containing cells were exposed to minimal medium without carbon source to trigger PHA degradation. Maximal PHA content in one set of experiments is set to 100% (knockout 11 w%; wild type 22 w%). The monomers 3-hydroxyoctanoate (white) and 3-hydroxyhexanoate (grey) are indicated.

A possible explanation for these observations might be the following: PHA accumulation and degradation might be a simultaneous process.[40, 41] In an intact cycle, part of precursors for PHA accumulation, i.e., acyl-CoA might result from previously depolymerized PHA. In the knockout mutant, CoA activation of monomers is blocked or hampered; hence recycling of monomers to form polymer is not possible. This might lead to the reduced PHA content in the knockout mutant. Assuming a cycle, simultaneous PHA accumulation by polymerization of CoA activated monomers during net PHA degradation would be possible in the wild type, but not in the mutant.

It has been recently reported that a *tesB*-like mutant of *Alcanivorax borkumensis* SK2 produces extracellular PHA.[308] *tesB* encodes an acyl-CoA thioesterase cleaving 3-hydroxyacyl-CoA thioesters bonds into CoA~SH and free 3-hydroxycarboxylic acids. Knockout of *tesB* can increase intracellular amount of 3-hydroxyacyl-CoA, leading to higher amount of PHA compared to the wild type.[308] The cyclic nature of PHA metabolism is not wasteful but rather of compensative character to guarantee balanced carbon and energy supply or redox potential within cells. The fate of PHA monomers was not clarified in the ACS1-negative mutant. Whether monomers can be degraded via different pathways or whether they are secreted into the supernatant, remains to be investigated. Export of fatty acids and probably also of their hydroxyl derivatives is achieved by either diffusion or by a process mediated by membrane-bound proteins. Since excess of fatty acids is toxic to cells,[297] it is more likely that they are excreted by certain mechanisms. In *Saccharomyces cerevisiae*, four *acs* genes were found and inactivation of one ACS resulted in accumulation of free fatty acids in the culture supernatant.[297] The attenuation of ACS activity might trigger an active export of fatty acids for cellular homeostasis.[297] A similar mechanism might be involved in ACS1-negative *P. putida* GPo1 and this could explain its lower accumulation and faster degradation of PHA.

Conclusion

The question was addressed whether or not a knockout of *acs1*, encoding one of two ACS in *P. putida* GPo1, would prevent degradation of PHA monomers and hence PHA accumulation would increase. We speculated that higher concentrations of PHA and PHA monomers might be reached by cultivation of this ACS1-knockout mutant, since recycling of PHA monomers would be impaired. However, it was not an effective tactic for the overproduction of mcl-PHA or their corresponding monomers.

Nevertheless, the mutant enabled insights into the underlying mechanisms of PHA metabolism. Compared to the parental strain, the knockout mutant accumulated less PHA and degradation of PHA occurred at a higher rate under carbon starvation conditions. Therefore, ACS1 seemed to play an important role not only for PHA degradation but also for its accumulation. Our results are in agreement with the concept of a cyclic nature of PHA metabolism, which accounts for lower PHA content in the mutant. The reason for faster PHA degradation still needs to be clarified.

It is not yet clear whether ACS2 can replace the function of ACS1. The prolonged lag phase observed with the mutant might indicate that ACS2 has to be induced first and then can assume the role of ACS1. Uptake and activation of fatty acids as carbon and energy source are not impaired in the mutant since similar growth rates and cell dry weight were reached when grown on octanoate as sole carbon source. Hence, ACS1 is not essential for functional β-oxidation.

In order to gain deeper insights into the role of ACS in PHA-metabolism, further investigation would be necessary; especially knockout mutants of the second ACS and *in vitro* studies on activity and substrate specificity with the corresponding purified enzymes are considered to be helpful.

Chapter 8

General Discussion

In this thesis HA from bacterial PHA regarding their biotechnological production as well as intracellular processes of their formation and degradation were investigated. An innovative bioprocess to provide HA as fine chemicals for further applications was presented. The method is straightforward, easy to handle and results in high yields (introduced in chapters 2-4). Following the observations of this study, attempts to scale up the process and potential applications of HA remain topics of research. The underlying mechanisms of PHA degradation were also addressed, including issues of physiological advantages of HA excretion for cells (chapter 5) and characterization of enzymes having HA as substrates (chapters 6 and 7). Even though some aspects of mcl-PHA metabolism were clarified for *P. putida* GPo1, many details about intracellular PHA degradation remain obscure and require further experiments and discussion. Knowledge about their molecular context in the bacterial cell is crucial for efficient PHA and HA production.

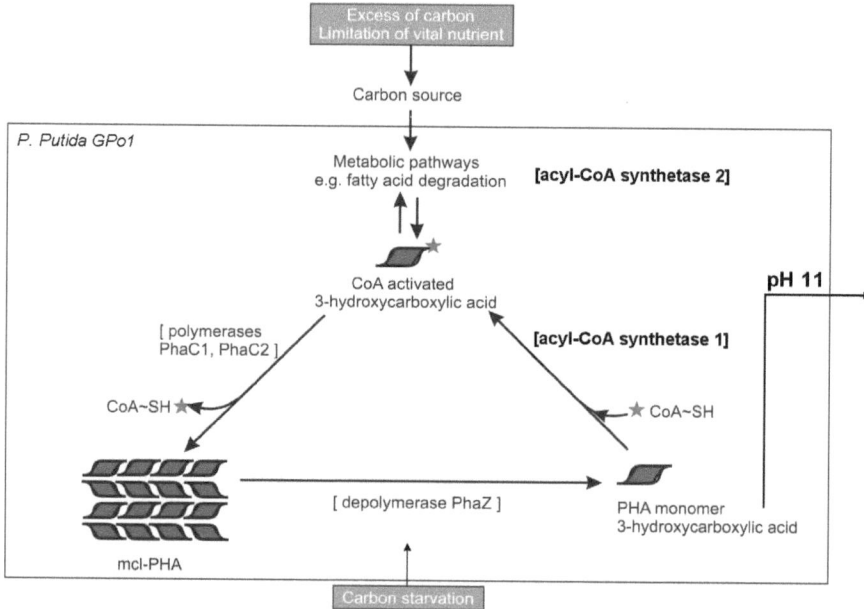

Figure 8.1 Main focus of investigations in this thesis: Metabolism of mcl-PHA includes a granule associated acyl-CoA synthetase (ACS1). A second acyl-CoA synthetase (ACS2) did not seem to be directly involved. *R*-enantiomers of HA were excreted at alkaline pH (maximum at pH 11) and this increased viability under such conditions.

Metabolic pathways

An excellent producer strain is important for efficient PHA production. PHA producer strains can further be improved by metabolic engineering, which often involves blocking pathways of PHA degradation or impairing pathways in competition with PHA production. For mcl-PHA, several strains have been investigated in this respect, one of which is a genetically engineered strain of *P. putida* U. Sandoval et al. described PHA overproduction in a knockout mutant *P. putida* U Δ*fadBA*.[224] Knocking out *fadBA* removes two enzymes of β-oxidation, a 3-hydroxyacyl-CoA dehydrogenase (FadB) and a ketothiolase (FadA). This blocked further β-oxidation of 3-hydroxyacyl-CoA intermediates, which are also precursors for PHA. Hence, PHA monomers and their precursors cannot be catabolized but can only be accumulated in the form of PHA (compare with Fig. 1.5). This led to accumulation of large amounts of aromatic PHA with more than 90% of cytoplasms being occupied by PHA granules.[224] The reason why PHA is accumulated and not any other intermediate might be the following: As long as a carbon source is in excess, regulatory systems favor PHA synthesis over PHA degradation, i.e., net PHA accumulation is observed.

In this thesis, the knockout mutant *P. putida* GPo1 *acs1::km* has been described, in which catabolism of PHA monomers was expected to be blocked. However, we observed fast degradation of PHA. One might speculate that ACS2 (acyl-CoA-synthetase 2) could replace the role of ACS1 or that the concentration of 3-hydroxyacyl-CoA, the PHA precursors, might be lower due to blocked cycle of PHA metabolism. Once more details about regulation and function of ACSs will be available, this knowledge might even allow constructing a novel PHA overproducers. For example, overexpression of ACS1 instead of its elimination might be favorable for PHA and HA production. Similar approaches have been reported; for example, acyl-CoA thioesterase in *Alcanivorax borkumensis* SK2 catalyzes the reverse reaction of ACS1, and blocking of this enzyme was already successfully used for PHA overproduction.[308]

Degradation pathways

The production of HA by directing the metabolic flux of intermediates has been investigated by several groups; however, the recovery of pure enantiomers remained challenging. For example, the mutant *P. putida* U Δ*fadBA* was further equipped with a plasmid to over-express PhaZ, which resulted in efficient excretion of racemic 3-hydroxy-*n*-phenylalkanoic acids.[224] Since PhaZ was constantly over-expressed, any PHA formed was instantly hydrolyzed. Consecutive enzymes comprise an acyl-CoA synthetase and an epimerase, but further reactions pathways are blocked in this mutant. This must result in an accumulation of the intermediates (*R*)- and (*S*)-3-hydroxyacyl-CoA and free (*R*)-3-hydroxycarboxylic acid. These compounds were excreted by some kind of overflow mechanism and a racemic mixture of 3-hydroxycarboxylic acids was found in the supernatant. This leads to the assumption that mainly CoA-activated compounds were channeled out of cells; however the mechanism is not clear at all. If free hydroxycarboxylic acids were released, a higher amount of *R*-enantiomers would have been found. Before leaving the cell, release and recycling of CoA has to occur because cells cannot afford to loose high amounts of this important cofactor. Whether or not PHA monomers are also excreted in *P. putida* Δ*fadBA* or in *P. putida phaZ* under conditions described in Sandoval et al. (2005), and which enantiomer is preferentially formed by those mutants has not been investigated. It has been reported that PHB depolymerase exhibits an intrinsic thiolase function[41], a fact that has not been confirmed for mcl-PHA depolymerase.[46] Potentially, the reaction product also depends on the most readily available reactant during conversion, thus the ratio of HS~CoA to water concentration might determine the product. Hence, a final conclusion cannot be drawn whether PHA monomers were released in their CoA-activated or free acid form.

Considerations on production of enantio pure compounds

Pure enantiomers were not obtained from *P. putida* U Δ*fadBA*, probably due to a 3-hydroxyacyl-CoA epimerase.[224] Only aromatic monomers and no aliphatic monomers were released in this mutant because it is able to further degrade aliphatic compounds via a second set of β-oxidation enzymes, which became active when FadA and FadB were missing.[224] These are two disadvantages when compared with the method presented in this thesis. Sandoval's approach is a simultaneous process,[224] whereas the HA production from PHA described in this thesis includes a first step of PHA accumulation and a second step of *in vivo* depolymerization and monomer release. This ensures that all precursors are bound as polymer (in pure *R*-configuration) prior to monomer production. *In vivo* depolymerization conditions (extracellular pH > 10) favor PHA depolymerase activity and may hamper acyl-CoA synthetase activity. Hence, high concentrations of free (*R*)-3-hydroxycarboxylic acids arise and leave cells by either diffusion or an active transport mechanism. In the supernatant HA showed absolute *R*-configuration. Why was not any trace of *S*-enantiomers detected? Enzymes of β-oxidation are organized in an enzyme complex and once a substrate is bound, it undergoes a multi-step conversion before being released. Any (*S*)-3-hydroxyacyl-CoA eventually formed during this process would immediately be reduced to 3-keto-acyl-CoA by 3-hydroxyacyl-CoA dehydrogenase (FadB) and then be further catabolized. These are fast reactions; the rate-limiting step during β-oxidation is the release of acetyl-CoA by ketothiolase (FadA).[309] In Sandoval's mutant, these reactions to dispose off eventually formed (*S*)-3-hydroxyacyl-CoA are blocked.

During studies with *P. putida,* only free HA were detected in the supernatant.[46, 191] The question remains whether free or CoA-activated HA is channeled to the cellular membrane. Uchino et al.[41] observed an intrinsic thiolase function of PHB depolymerase from *C. necator*. Hence, they claim that CoA-activated monomers are released from PHA. In contrast to this, de Eugenio et al.[46] observed production of free HA by purified PhaZ from *P. putida* KT2440. It might be speculated that both water and CoA-SH can serve as nucleophil for polyester hydrolyzation in the same enzyme under appropriate conditions.

In vivo depolymerization experiments with an ACS1 knockout in *P. putida* GPo1 might elucidate whether or not free or CoA-activated HA are overproduced. If free HA are accumulated, wild type and mutant would show similar phenotypes with regard to monomer excretion. If the overflow pool arises from CoA-activated HA, a mutant with blocked ACS would not be able to release PHA monomers. However, *P. putida* GPo1 contains two types of ACS that are likely to complement each other and a double knockout mutant of both ACS is thought to not be capable to survive. Thus, the actual mecha-

Insights into regulation of PHA metabolism

A sophisticated regulatory system links environmental conditions, such as nutrient supply, to expression of genes, which control PHA metabolism; however the mechanism is not yet fully understood. The influence of granule-associated proteins also seems to be crucial. Besides phasins, PHA polymerases and PHA depolymerases, also ACS1 was shown to be localized at PHA granules. This might indicate that ACS could have a regulatory function as well, similar to phasins. During last decades a lot of experimental data have been accumulated to allow insight on PHA metabolism in *P. putida*. The regulatory network of PHA metabolism includes transcriptional and enzymatic factors and its understanding can assist to maximize PHA production. Functional analyses of involved proteins were accomplished based on a *pha locus* knockout mutant of *P. putida U*.[310] The mutant is unable to synthesize PHA and bacterial cells are smaller, indicating that the deletion causes a loss in size and volume control. Subsequent introduction of *pha* genes unraveled their particular function, which might have important biotechnological implication for establishing improved polymers.[310] Expression of *phaC1* is sufficient to ensure synthesis of aromatic and aliphatic PHA and *phaZ* is required for its hydrolysis, however proper modulation of number and shape of granules needs further proteins. Co-expression of PhaC1 and phasin PhaF restore normal granule architecture. In the absence of PhaF aromatic PHA can only be synthesized when phaC1 is over-expressed, indicating that aromatic precursors with a more rigid structure are worse substrates for PhaC1 and PhaF can facilitate their polymerization, probably by binding of PhaF to the nascent polymer chain. For example, PHA over-producer mutant *P. putida* U ΔfadBA[248] equipped with *phaF* led to giant cells with >95% of the cytoplasm occupied by PHA; these cells finally lyzed and released accumulated PHA. PhaI is a repressor of phaF, preventing exorbitant increase of cellular volume; and *phaD* acts as a transcriptional activator of phaF by competing with PhaI for the P*f* promoter.[310] Attempts to incorporate new monomeric units might be designed more successfully, taking these mechanisms into account. Considering the relatively low substrate specificity of PHA polymerases, fine tuning of the expression of *phaFDI*, and eventually also of *acs1*, may allow the formation of PHA with many interesting monomers.

The regulatory network of stress tolerance seems to also be interconnected with PHA accumulation and degradation.[51, 272] Excess of a carbon source with simultaneous limitation of vital nutrients triggers PHA formation.[21] PHA degradation upon pH shift to an alkaline range (chapter 5), as well as the observation of increased ppGpp and RpoS levels when PHA is hydrolyzed, implicate integration of PHA metabolism in the cellular stress response network.[159, 311] Therefore, knowledge of these factors might be advantageous to optimize a production process.

The production not only of HA, but also of the corresponding dimers, trimers or other desired building blocks might be of future interest. Different methods could be applied for this purpose including site-directed mutagenesis, or directed evolution of involved enzymes, or overexpression of required enzymes in other hosts. Genetic engineering of *phaC* or *phaZ* or *acs* might offer ways to increase or decrease substrate specificity and thereby achieve desired polymer/monomer properties. Zheng et al. reported that N-terminal mutations of PhaC from *C. necator* changed the enzyme's activity and specificity.[312] The formation of dimers was observed with PhaZ from *P. putida* KT2442, for instance.[46] Hence, exploitation of intrinsic properties and sophisticated modulation of PHA synthesizing and degrading enzymes could make the formation of many chiral compounds derived from PHA accessible.

Future of PHA and related compounds

Plastic waste is a huge ecological problem and it is accepted that avoiding waste production is more sustainable than its recycling. Furthermore, crude oil reserves are considered to be depleted to 90% within the next 10-20 years.[313] Market for biodegradable biomaterial could reach a volume of 10 bio US dollars in 2010.[314] Bioplastics still have to compete with classical polymers like PP, PS, or PE, which cost <1 US-dollar per kilogram. Even though most bioplastics are still more expensive (than, e.g., PET[315]), the price relationship might shift when crude oil becomes scarce and ecological awareness of the society increases further.

Efficient production of HA

Many soil bacteria are able to produce PHA and many of them grow on contaminated soil, e.g., containing hydrocarbons. For instance, *P. putida* GPo1 can grow on pollutants like hydrocarbons.[178] Combination of HA production with bioremediation of polluted soil could result in an environmentally friendly and value generating process by, e.g., using a PHA producer strain such as *P. putida* GPo1 and extract secreted HA from the supernatant.

PHA and HA could be produced in a concerted process which keeps expenses at a minimum and renders their biosynthesis even more profitable and efficient. Currently, most homogenous and quality controlled PHA can be synthesized using continuous cultivation. Reproducibility and homogenously composed PHA requires chemostat conditions and at initial production phases or after switching parameters these demands are not fulfilled. Hence, these PHA-containing cells are normally discarded but could as well be directly used for HA production. Therefore, if PHA was produced by continuous cultivation in a chemostat, cells harvested prior to steady-state conditions could be used to generate HA.

Cost reduction is essential for economic production of HA. The limiting step within the described procedure is separation of HA by chromatography. Larger amounts could be produced by applying chromatography columns with extended size and diameter; hence scale-up for synthesizing several hundred grams is feasible. Since recycling of solvents and column packing material is possible, waste formation can be kept at a minimum.

Potential applications

Potential applications of HA have been discussed in the introduction. For example, they could serve as synthons for the design of new antibiotics and fungicides. Some HA inhibited growth of several *Listeria* species in a mM range (chapter 2). When coupling short peptides to HA (here: Lys-Leu-Leu-Lys plus (R)-3-hydroxynon-8-enoic acid) we observed even increased antimicrobial activity (data not shown). Even though concentrations are still high when compared with established antibiotics, research in this direction nevertheless seems promising.

Besides many other potential applications, HA could be used as monomers to chemically synthesize homopolymers and tailor-made polymers. The possibility to create new types of polyesters and lower endotoxin contents would be additional benefits. This is ensured by *in vivo* depolymerization, a process to recover HA in which bacterial cells are not destroyed. In contrast to this, the recovery of bacterial PHA from lysed cells is currently the method of choice and implicates dealing with a certain endotoxin concentration.

Biotechnologically produced compounds engage a great market volume (50 billion US dollar) and involved industries are expected to grow during the next years.[316] Furthermore, there is political and public concern to improve industrial sustainability. In this context, chiral HA have prospects as synthons for pharmaceuticals and fine chemicals and their intrinsic antimicrobial properties as well as their environmentally friendly production process are additional advantages.

There are currently many challenges for our society and research in biotechnology and environmental science seems to provide quite promising approaches to solve urgent problems. Therefore, we believe that research topics addressed in this thesis will be a small contribution to a more sustainable world.

References

1. Berton, J., Continent-size toxic stew of plastic trash fouling swath of Pacific Ocean. *San Francisco Chronicle* **2007**, W8.
2. Moore, C. J.; Moore, S. L.; Leecaster, M. K.; Weisberg, S. B., A comparison of plastic and plankton in the North Pacific central gyre. *Mar. Pollut. Bull.* **2001**, 42, (12), 1297-1300.
3. Ward, A. J. W.; Webster, M. M.; Hart, P. J. B., Social recognition in wild fish populations. *Proc. R. Soc. London* **2007**, B, (274), 1071-1077.
4. Bommarius, A. S.; Riebel, B. R., *Biocatalysis - Fundamentals and applications*. 1 ed.; Wiley-VCH: Weinheim, 2004; p 611.
5. Kohler, H.-P. E.; Angst, W.; Giger, W.; Kanz, C.; Müller, S.; Suter, M. J.-F., Environmental fate of chiral pollutants - the necessity of considering stereochemistry. *Chimia* **1997**, 51, 947-951.
6. Kohler, H.-P. E.; Nickel, K.; Zipper, C., Effect of chirality on the microbial degradation and the environmental fate of chiral pollutants. *Adv. Microb. Ecol.* **2000**, 16, 201-231.
7. Sheldrake, G. N.; Crosby, J., *Chirality in industry: The commercial manufacture and applications of optically active compounds*. John Wiley & Sons: New York, 1998; p 426.
8. Gavrilescu, M.; Chisti, Y., Biotechnology - A sustainable alternative for chemical industry. *Biotechnol. Adv.* **2005**, 23, (7-8), 471-499.
9. Sheldon, R. A.; van Rantwijk, F., Biocatalysis for sustainable organic synthesis. *Aust. J. Chem.* **2004**, 57, (4), 281-289.
10. Guerittevoegelein, F.; Guenard, D.; Lavelle, F.; Legoff, M. T.; Mangatal, L.; Potier, P., Relationships between the structure of taxol analogs and their antimitotic activity. *J. Med. Chem.* **1991**, 34, (3), 992-998.
11. Schmid, A.; Dordick, J. S.; Hauer, B.; Kiener, A.; Wubbolts, M.; Witholt, B., Industrial biocatalysis today and tomorrow. *Nature* **2001**, 409, (6817), 258-268.
12. Kim, Y. B.; Lenz, R. W., Polyesters from microorganisms. In *Advances in biochemical engineering/biotechnology*, Babel, W.; Steinbüchel, A., Eds. Springer-Verlag: Berlin Heidelberg, 2001; Vol. 71, pp 51-79.
13. Steinbüchel, A.; Valentin, H. E., Diversity of bacterial polyhydroxyalkanoic acids. *FEMS Microbiol. Lett.* **1995**, 128, (3), 219-228.
14. Zinn, M.; Witholt, B.; Egli, T., Occurrence, synthesis and medical application of bacterial polyhydroxyalkanoate. *Adv. Drug Del. Rev.* **2001**, 53, (1), 5-21.
15. Jendrossek, D.; Handrick, R., Microbial degradation of polyhydroxyalkanoates. *Annu. Rev. Microbiol.* **2002**, 56, 403-432.
16. Leaf, T. A.; Peterson, M. S.; Stoup, S. K.; Somers, D.; Srienc, F., *Saccharomyces cerevisiae* expressing bacterial polyhydroxybutyrate synthase produces poly-3-hydroxybutyrate. *Microbiology* **1996**, 142, 1169-1180.
17. Huisman, G. W.; Wonink, E.; Meima, R.; Kazemier, B.; Terpstra, P.; Witholt, B., Metabolism of poly(3-hydroxyalkanoates) (PHAs) by *Pseudomonas oleovorans* - Identification and sequences of genes and function of the encoded proteins in the synthesis and degradation of PHA. *J. Biol. Chem.* **1991**, 266, (4), 2191-2198.
18. Williams, M. D.; Fieno, A. M.; Grant, R. A.; Sherman, D. H., Expression and analysis of a bacterial poly(hydroxyalkanoate) synthase in insect cells using a baculovirus system. *Protein Expression Purif.* **1996**, 7, (2), 203-211.
19. Poirier, Y.; Nawrath, C.; Somerville, C., Production of polyhydroxyalkanoates, a family of biodegradable plastics and elastomers, in bacteria and plants. *Biotechnology* **1995**, 13, (2), 142-150.
20. Du, G.; Yu, J., Green technology for conversion of food scraps to biodegradable thermoplastic polyhydroxyalkanoates. *Environ. Sci. Technol.* **2002**, 36, (24), 5511-5516.
21. Anderson, A. J.; Dawes, E. A., Occurence, metabolism, metabolic role, and industrial uses of bacterial polyhydroxyalkanoates. *Microbiol. Rev.* **1990**, 54, 450-472.

22. Lenz, R. W.; Marchessault, R. H., Bacterial polyesters: Biosynthesis, biodegradable plastics and biotechnology. *Biomacromolecules* **2005**, 6, (1), 1-8.

23. Wallen, L. L.; Rohwedde, W. K., Poly-β-hydroxyalkanoate from activated sludge. *Environ. Sci. Technol.* **1974**, 8, (6), 576-579.

24. Rehm, B. H. A., Biogenesis of microbial polyhydroxyalkanoate granules: A platform technology for the production of tailor-made bioparticles. *Curr. Issues Mol. Biol.* **2007**, 9, 41-62.

25. Byrom, D., Polymer synthesis by microorganisms - Technology and economics. *Trends Biotechnol.* **1987**, 5, (9), 246-250.

26. Korherr, C.; Roth, M.; Holler, E., Poly(β-L-malate) hydrolase from plasmodia of *Physarum polycephalum*. *Can. J. Microbiol.* **1995**, 41, 192-199.

27. Elbanna, K.; Lutke-Eversloh, T.; Jendrossek, D.; Luftmann, H.; Steinbuchel, A., Studies on the biodegradability of polythioester copolymers and homopolymers by polyhydroxyalkanoate (PHA)-degrading bacteria and PHA depolymerases. **2004**, 182, (2-3), 212-225.

28. Lutke-Eversloh, T.; Bergander, K.; Luftmann, H.; Steinbuchel, A., Identification of a new class of biopolymer: bacterial synthesis of a sulfur-containing polymer with thioester linkages. **2001**, 147, 11-19.

29. Doi, Y.; Tamaki, A.; Kunioka, M.; Soga, K., Biosynthesis of terpolyesters of 3-hydroxybutyrate, 3-hydroxyvalerate, and 5-hydroxyvalerate in *Alcaligenes eutrophus* from 5-chloropentanoic and pentanoic acids. *Macromol. Rapid Commun.* **1987**, 8, (12), 631-635.

30. Füchtenbusch, B.; Fabritius, D.; Wältermann, M.; Steinbüchel, A., Biosynthesis of novel copolyesters containing 3-hydroxypivalic acid by *Rhodococcus ruber* NCIMB 40126 and related bacteria. *FEMS Microbiol. Lett.* **1998**, 159, (1), 85-92.

31. Hartmann, R.; Hany, R.; Geiger, T.; Egli, T.; Witholt, B.; Zinn, M., Tailored biosynthesis of olefinic medium-chain-length poly [(R)-3-hydroxyalkanoates] in *Pseudomonas putida* GPo1 with improved thermal properties. *Macromolecules* **2004**, 37, (18), 6780-6785.

32. Lee, S. Y., Bacterial polyhydroxyalkanoates. *Biotechnol. Bioeng.* **1996**, 49, (1), 1-14.

33. Steinbüchel, A.; Hein, S., Biochemical and molecular basis of microbial synthesis of polyhdroxyalkanoates in microorganisms. In *Advances in biochemical engineering/biotechnology*, Babel, W.; Steinbüchel, A., Eds. Springer-Verlag: Berlin Heidelberg, 2001; Vol. 71, pp 81-123.

34. Schubert, P.; Krüger, N.; Steinbüchel, A., Molecular analysis of the *Alcaligenes eutrophus* poly(3-hydroxybutyrate) biosynthetic operon - Identification of the N-terminus of poly(3-hydroxybutyrate) synthase and identification of the promoter. *J. Bacteriol.* **1991**, 173, (1), 168-175.

35. Pohlmann, A.; Fricke, W. F.; Reinecke, F.; Kusian, B.; Liesegang, H.; Cramm, R.; Eitinger, T.; Ewering, C.; Potter, M.; Schwartz, E.; Strittmatter, A.; Voss, I.; Gottschalk, G.; Steinbüchel, A.; Friedrich, B.; Bowien, B., Genome sequence of the bioplastic-producing "Knallgas" bacterium *Ralstonia eutropha* H16. *Nat. Biotechnol.* **2006**, 24, (10), 1257-1262.

36. Madison, L. L.; Huisman, G. W., Metabolic engineering of poly(3-hydroxyalkanoates): From DNA to plastic. *Microb. Mol. Biol. Rev.* **1999**, 63, (1), 21-53.

37. Prieto, M. A.; Bühler, B.; Jung, K.; Witholt, B.; Kessler, B., PhaF, a polyhydroxyalkanoate-granule-associated protein of *Pseudomonas oleovorans* GPo1 involved in the regulatory expression system for pha genes. *J. Bacteriol.* **1999**, 181, (3), 858-868.

38. Garcia, B.; Olivera, E. R.; Minambres, B.; Fernandez-Valverde, M.; Canedo, L. M.; Prieto, M. A.; Garcia, J. L.; Martinez, M.; Luengo, J. M., Novel biodegradable aromatic plastics from a bacterial source - Genetic and biochemical studies on a route of the phenylacetyl-CoA catabolon. *J. Biol. Chem.* **1999**, 274, (41), 29228-29241.

39. Taidi, B.; Mansfield, D. A.; Anderson, A. J., Turnover of poly(3-hydroxybutyrate) (PHB) and its influence on the molecular mass of the polymer accumulated by *Alcaligenes eutrophus* during batch culture. *FEMS Microbiol. Lett.* **1995**, 129, (2-3), 201-205.

40. Doi, Y.; Segawa, A.; Kawaguchi, Y.; Kunioka, M., Cyclic nature of poly(3-hydroxyalkanoate) metabolism in *Alcaligenes eutrophus*. *FEMS Microbiol. Lett.* **1990**, 67, (1-2), 165-169.

References

41. Uchino, K.; Saito, T.; Gebauer, B.; Jendrossek, D., Isolated poly(3-hydroxybutyrate) (PHB) granules are complex bacterial organelles catalyzing formation of PHB from acetyl coenzyme A (CoA) and degradation of PHB to acetyl-CoA. *J. Bacteriol.* **2007**, 189, (22), 8250–8256.

42. Senior, P. J.; Dawes, E. A., Regulation of poly-β-hydroxybutyrate metabolism in *Azotobacter beijerinckii*. *Biochem. J.* **1973**, 134, (1), 225-238.

43. York, G. M.; Junker, B. H.; Stubbe, J.; Sinskey, A. J., Accumulation of the PhaP phasin of *Ralstonia eutropha* is dependent on production of polyhydroxybutyrate in cells. *J. Bacteriol.* **2001**, 183, (14), 4217-4226.

44. Oeding, V.; Schlegel, J. G., ß-Ketohthiolase from *Hydrogenomonas eutropha* H16 and its significance in the regulation of poly-ß-hydroxybutyrate metabolism. *Biochem. J.* **1973**, 134, 239-248.

45. Haywood, G. W.; Anderson, A. J.; Chu, L.; Dawes, E. A., Accumulation of polyhydroxyalkanoates by bacteria and the substrate specificities of the biosynthetic enzymes. *Biochem. Soc. Trans.* **1988**, 16, (6), 1046-1047.

46. de Eugenio, L. I.; Garcia, P.; Luengo, J. M.; Sanz, J. M.; San Roman, J.; Garcia, J. L.; Prieto, M. A., Biochemical evidence that *phaZ* gene encodes a specific intracellular medium-chain-length polyhydroxyalkanoate depolymerase in *Pseudomonas putida* KT2442 - Characterization of a paradigmatic enzyme. *J. Biol. Chem.* **2007**, 282, (7), 4951-4962.

47. O'Leary, N. D.; O'Connor, K. E.; Ward, P.; Goff, M.; Dobson, A. D. W., Genetic characterization of accumulation of polyhydroxyalkanoate from styrene in *Pseudomonas putida* CA-3. *Appl. Environ. Microbiol.* **2005**, 71, (8), 4380-4387.

48. Jendrossek, D., Microbial degradation of polyesters. In *Advances in biochemical engineering/biotechnology*, Babel, W.; Steinbüchel, A., Eds. Springer-Verlag: Berlin Heidelberg, 2001; Vol. 71, pp 293-325.

49. Timm, A.; Steinbüchel, A., Cloning and molecular analysis of the poly(3-hydroxyalkanoic acid) gene locus of *Pseudomonas aeruginosa* Pao1. *Eur. J. Biochem.* **1992**, 209, (1), 15-30.

50. Hoffmann, N.; Rehm, B. H. A., Regulation of polyhydroxyalkanoate biosynthesis in *Pseudomonas putida* and *Pseudomonas aeruginosa*. *FEMS Microbiol. Lett.* **2004**, 237, (1), 1-7.

51. Kessler, B.; Witholt, B., Factors involved in the regulatory network of polyhydroxyalkanoate metabolism. *J. Biotechnol.* **2001**, 86, (2), 97-104.

52. Qi, Q.; Steinbüchel, A.; Rehm, B. H. A., In vitro synthesis of poly(3-hydroxydecanoate): Purification and enzymatic characterization of type II polyhydroxyalkanoate synthases PhaC1 and PhaC2 from *Pseudomonas aeruginosa*. *Appl. Microbiol. Biotechnol.* **2000**, 54, (1), 37-43.

53. Zinn, M.; Witholt, B.; Egli, T., Dual nutrient limited growth: Models, experimental observations, and applications. *J. Biotechnol.* **2004**, 113, 263-279.

54. Monod, J., The growth of bacterial cultures. *Ann. Rev. Microbiol.* **1949**, 3, 371-394.

55. Pirt, S. J., *Principles of microbe and cell cultivation*. Blackwell: London, 1975; p 274.

56. Kellerhals, M. B.; Kessler, B.; Witholt, B., Closed-loop control of bacterial high-cell-density fed-batch cultures: Production of mcl-PHAs by *Pseudomonas putida* KT2442 under single-substrate and cofeeding conditions. *Biotechnol. Bioeng.* **1999**, 65, (3), 306-315.

57. Lee, E. Y.; Choi, C. Y., Gas-chromatography mass-spectrometric analysis and its application to a screening-procedure for novel bacterial polyhydroxyalkanoic acids containing long-chain saturated and unsaturated monomers. *J. Ferment. Bioeng.* **1995**, 80, (4), 408-414.

58. Suzuki, T.; Yamane, T.; Shimizu, S., Kinetics and effect of nitrogen-source feeding on production of poly-β-hydroxybutyric acid by fed-batch culture. *Appl. Microbiol. Biotechnol.* **1986**, 24, (5), 366-369.

59. Majid, M. I. A.; Akmal, D. H.; Few, L. L.; Agustien, A.; Toh, M. S.; Samian, M. R.; Najimudin, N.; Azizan, M. N., Production of poly(3-hydroxybutyrate) and its copolymer poly(3-hydroxybutyrate-co-3-hydroxyvalerate) by *Erwinia sp* USMI-20. *Int. J. Biol. Macromol.* **1999**, 25, (1-3), 95-104.

60. Lee, S. Y.; Choi, J. I.; Wong, H. H., Recent advances in polyhydroxyalkanoate production by bacterial fermentation: mini-review. **1999**, 25, (1-3), 31-36.

61. Zinn, M. Dual (C,N) nutrient limited growth of *Pseudomonas oleovorans*. PhD thesis, ETH Zürich, Zürich, 1998.

62. Durner, R.; Witholt, B.; Egli, T., Accumulation of poly[(R)-3-hydroxyalkanoates] in *Pseudomonas oleovorans* during growth with octanoate in continuous culture at different dilution rates. *Appl. Environ. Microbiol.* **2000**, 66, (8), 3408-3414.

63. Durner, R.; Zinn, M.; Witholt, B.; Egli, T., Accumulation of poly[(R)-3-hydroxyalkanoates] in *Pseudomonas oleovorans* during growth in batch and chemostat culture with different carbon sources. *Biotechnol. Bioeng.* **2001**, 72, (3), 278-288.

64. Egli, T.; Zinn, M., The concept of multiple-nutrient-limited growth of micro-organisms and its application in biotechnological processes. *Biotechnol. Adv.* **2003**, 22, 35-43.

65. Egli, T., On multiple-nutrient-limited growth of microorganisms, with special reference to dual limitation by carbon and nitrogen substrates. *A. v. Leeuwenhoek* **1991**, 60, 225-234.

66. Poirier, Y.; Somerville, C.; Schechtman, L. A.; Satkowski, M. M.; Noda, I., Synthesis of high-molecular-weight poly([R]-(-)-3- hydroxybutyrate) in transgenic *Arabidopsis thaliana* plant cells. *Int. J. Biol. Macromol.* **1995**, 17, (1), 7-12.

67. Kellerhals, M. B.; Hazenberg, W.; Witholt, B., High cell density fermentations of *Pseudomonas oleovorans* for the production of mcl-PHAs in two-liquid phase media. *Enzyme Microb. Technol.* **1999**, 24, (1-2), 111-116.

68. Page, W. J., Bacterial polyhydroxyalkanoates, natural biodegradable plastics with a great future. *Can. J. Microbiol.* **1995**, 41, (Suppl. 1), 1-3.

69. Williams, S. F.; Martin, D. P.; Horowitz, D. M.; Peoples, O. P., PHA applications: Addressing the price performance issue. *Int. J. Biol. Macromol.* **1999**, 25, (1-3), 111-121.

70. Lee, S. Y.; Lee, Y.; Wang, F. L., Chiral compounds from bacterial polyesters: Sugars to plastics to fine chemicals. *Biotechnol. Bioeng.* **1999**, 65, (3), 363-368.

71. Marois, Y.; Zhang, Z.; Vert, M.; Deng, X. Y.; Lenz, R.; Guidoin, R., Hydrolytic and enzymatic incubation of polyhydroxyoctanoate (PHO): A short-term in vitro study of a degradable bacterial polyester. *J. Biomater. Sci., Polym. Ed.* **1999**, 10, (4), 483-499.

72. Pouton, C. W.; Akhtar, S., Biosynthetic polyhydroxyalkanoates and their potential in drug delivery. *Adv. Drug Del. Rev.* **1996**, 18, (2), 133-162.

73. Sendil, D.; Gursel, I.; Wise, D. L.; Hasirci, V., Antibiotic release from biodegradable PHBV microparticles. *J. Control. Rel.* **1999**, 59, (2), 207-217.

74. Sodian, R.; Sperling, J. S.; Martin, D. P.; Egozy, A.; Stock, U.; Mayer, J. E.; Vacanti, J. P., Fabrication of a trileaflet heart valve scaffold from a polyhydroxyalkanoate biopolyester for use in tissue engineering. *Tissue Eng.* **2000**, 6, (2), 183-188.

75. de Koning, G. J. M., Physical properties of bacterial poly((R)-3-hydroxyalkanoates). *Can. J. Microbiol.* **1995**, 41, (Suppl. 1), 303-309.

76. Arslan, H.; Mentes, A.; Hazer, B., Synthesis and characterization of diblock, triblock, and multiblock copolymers containing poly(3-hydroxy butyrate) units. *J. Appl. Polym. Sci.* **2004**, 94, (4), 1789-1796.

77. Haywood, G. W.; Anderson, A. J.; Dawes, E. A., The importance of PHB-synthase substrate specificity in polyhydroxyalkanoate synthesis by *Alcaligenes eutrophus*. *FEMS Microbiol. Lett.* **1989**, 57, 1-6.

78. Sudesh, K.; Abe, H.; Doi, Y., Synthesis, structure and properties of polyhydroxyalkanoates: Biological polyesters. *Prog. Polym. Sci.* **2000**, 25, (10), 1503-1555.

79. Haywood, G. W.; Anderson, A. J.; Chu, L.; Dawes, E. A., Characterization of two 3-ketothiolases possessing differing substrate specificities in the polyhydroxyalkanoate synthesizing organism *Alcaligenes eutrophus*. *FEMS Microbiol. Lett.* **1988**, 52, 91-96.

80. Lee, S. Y.; Park, S. H.; Lee, Y.; Lee, S. H., Production of chiral and other valuable compounds from microbial polyester. In *Biopolymers*, Doi, Y.; Steinbüchel, A., Eds. Wiley-VCH Verlag GmbH: Weinheim, 2002; Vol. 4, pp 375-387.

81. Ohashi, T.; Hasegawa, J., New preparative methods for optically active β-hydroxycarboxylic acids. In *Chirality in Industry*, Collins, A. N.; Sheldrake, G. N.; Crosby, J., Eds. John Wiley & Sons: New York, 1992; Vol. 1, pp 249-268.

References

82. Brown, H. C.; Ramachandran, P. V., The boron approach to asymmetric synthesis. *Pure Appl. Chem.* **1991**, 63, (3), 307-316.

83. Noyori, R.; Ohkuma, T.; Kitamura, M.; Takaya, H.; Sayo, N.; Kumobayashi, H.; Akutagawa, S., Asymmetric hydrogenation of β-keto carboxylic esters - A practical, purely chemical access to β-hydroxy esters in high enantiomeric purity. *J. Am. Chem. Soc.* **1987**, 109, (19), 5856-5858.

84. Noyori, R.; Kitamura, M.; Ohkuma, T., Toward efficient asymmetric hydrogenation: Architectural and functional engineering of chiral molecular catalysts. *Proc. Natl. Acad. Sci. USA* **2004**, 101, (15), 5356-5362.

85. Ikunaka, M., A process in need is a process indeed: Scalable enantioselective synthesis of chiral compounds for the pharmaceutical industry. *Chem. Eur. J.* **2003**, 9, (2), 379-388.

86. Nakahata, M.; Imaida, M.; Ozaki, H.; Harada, T.; Tai, A., The preparation of optically pure 3-hydroxyalkanoic acid - The enantioface-differentiating hydrogenation of the C=O double-bond with modified Raney-nickel. *Bull. Chem. Soc. Jpn.* **1982**, 55, (7), 2186-2189.

87. Wang, Z.; Zhao, C.; Pierce, M. E.; Fortunak, J. M., Enantioselective synthesis of β-hydroxycarboxylic acids: Direct conversion of β-oxocarboxylic acids to enantiomerically enriched β-hydroxycarboxylic acids via neighboring group control. *Tetrahedron: Asymmetry* **1999**, 10, 225–228.

88. Zhang, J.; Duetz, W. A.; Witholt, B.; Li, Z., Rapid identification of new bacterial alcohol dehydrogenases for (R)- and (S)-enantioselective reduction of β-ketoesters. *Chem. Commun.* **2004**, 18, 2120-2121.

89. Zheng, Z.; Gong, Q.; Liu, T.; Deng, Y.; Chen, J. C.; Chen, G. Q., Thioesterase II of *Escherichia coli* plays an important role in 3-hydroxydecanoic acid production. *Appl. Environ. Microbiol.* **2004**, 70, (7), 3807-3813.

90. Sih, C. J.; Zhou, B. N.; Gopalan, A. S.; Shieh, W. R.; Chen, C. S.; Girdaukas, G.; Vanmiddlesworth, F., Enantioselective reductions of β-keto-esters by baker's yeast. *Ann. Ny. Acad. Sci.* **1984**, 434, (Dec), 186-193.

91. Utaka, M.; Watabu, H.; Higashi, H.; Sakai, T.; Tsuboi, S.; Torii, S., Asymmetric reduction of aliphatic short-chain to long-chain β-keto acids by use of fermenting baker's yeast. *J. Org. Chem.* **1990**, 55, (12), 3917-3921.

92. de Roo, G.; Kellerhals, M. B.; Ren, Q.; Witholt, B.; Kessler, B., Production of chiral R-3-hydroxyalkanoic acids and R-3-hydroxyalkanoic acid methylesters via hydrolytic degradation of polyhydroxyalkanoate synthesized by pseudomonads. *Biotechnol. Bioeng.* **2002**, 77, (6), 717-722.

93. Seebach, D.; Beck, A. K.; Breitschuh, R.; Job, K., Direct degradation of the biopolymer poly[(R)-3-hydroxybutyric acid] to (R)-3-hydroxybutanoic acid and its ester. *Org. Synth.* **1993**, 71, 39.

94. Foster, L. J. R.; Fuller, R. C.; Lenz, R. W., Activities of extracellular and intracellular depolymerases of polyhydroxyalkanoates. In *Hydrogels and biodegradable polymers for bioapplications*, Ottenbrite, R. M.; Huang, S. J.; Park, K., Eds. Oxford University Press: Oxford, 1996; Vol. 627, pp 68-92.

95. Foster, L. J. R.; Lenz, R. W.; Fuller, R. C., Intracellular depolymerase activity in isolated inclusion bodies containing polyhydroxyalkanoates with long alkyl and functional substituents in the side chain. *Int. J. Biol. Macromol.* **1999**, 26, (2-3), 187-192.

96. Wang, L.; Armbruster, W.; Jendrossek, D., Production of medium-chain-length hydroxyalkanoic acids from Pseudomonas putida in pH stat. **2007**, 75, (5), 1047-1053.

97. Steinbüchel, A., Polyhydroxyalkanoic acids. In *Biomaterials*, Byrom, D., Ed. Macmillan Publishers Ltd: Basingstoke, 1991; pp 123-213.

98. Chen, G.-Q.; Wu, Q., Microbial production and applications of chiral hydroxyalkanoates. *Appl. Microbiol. Biotechnol.* **2005**, 67, (5), 592-599.

99. Dedier, S.; Krebs, S.; Lamas, J. R.; Poenaru, S.; Folkers, G.; de Castro, J. A. L.; Seebach, D.; Rognan, D., Structure-based design of nonnatural ligands for the HLA-B27 protein. *J. Recept. Signal Transduction* **1999**, 19, (1-4), 645-657.

100. Tasaki, O.; Hiraide, A.; Shiozaki, T.; Yamamura, H.; Ninomiya, N.; Sugimoto, H., The dimer and trimer of 3-hydroxybutyrate oligomer as a precursor of ketone bodies for nutritional care. *J. Parenter. Enteral. Nutr.* **1999**, 23, (6), 321-325.

101. Abe, S., Massoy oil. *J. Chem. Soc. Jpn.* **1937**, 58, 246-251.

102. Kaiser, R.; Lamparsky, D., Das Lacton der 5-hydroxy-cis-2,cis-7-decadiensäure und weitere Lactone aus dem Absolue der Blüten von *Polianthes tuberosa* L. *Tetrahedron Lett.* **1976**, 17, (20), 1659-1660.

103. Touati, R.; Ratovelomanana-Vidal, V.; Ben-Hassine, B.; Genet, J. P., Synthesis of enantiopure (R)-(-)-massoialactone through ruthenium-SYNPHOS (R) asymmetric hydrogenation. *Tetrahedron: Asymmetry* **2006**, 17, (24), 3400-3405.

104. Satô, T., Synthesis of optically active forms of the δ-lactone of 3,5-dihydroxydecanoic acid. *Can. J. Chem.* **1987**, 65, (12), 2732-2733.

105. Schreiber, S. L.; Kelly, S. E.; Porco, J. A.; Sammakia, T.; Suh, E. M., Structural and synthetic studies of the spore germination autoinhibitor gloeosporone. *J. Am. Chem. Soc.* **1988**, 110, (18), 6210-6218.

106. Graner, G.; Hamberg, M.; Meijer, J., Screening of oxylipins for control of oilseed rape (*Brassica napus*) fungal pathogens. *Phytochemistry* **2003**, 63, (1), 89-95.

107. Hamberg, M., Hidden stereospecificity in the biosynthesis of divinyl ether fatty acids. *FEBS J.* **2005**, 272, (3), 736-743.

108. Marotta, E.; Pagani, I.; Righi, P.; Rosini, G., Synthesis of methyl-substituted bicyclo[3.2.0]hept-3-en-6-ones and 3,3a,4,6a-tetrahydro-2H-cyclopenta[b]furan-2-ones. *Tetrahedron* **1994**, 50, (25), 7645-7656.

109. Marotta, E.; Righi, P.; Rosini, G., A new effective route to methyl-substituted 3,3a,4,6a-tetrahydro-2H-cyclopenta[b]furan-2-ones. *Tetrahedron Lett.* **1994**, 35, (18), 2949-2950.

110. Hayakawa, K.; Nagatsugi, F.; Kanematsu, K., Total synthesis of (+)-4-oxo-5,6,9,10-tetradehydro-4,5-secofuranoeremophilane-5,1-carbolactone via novel lactone construction through allene intramolecular cyclo addition. *J. Org. Chem.* **1988**, 53, (4), 860-863.

111. Zhang, H.; Liao, Z. X.; Yue, J. M., Five new sesquiterpenoids from *Parasenecio petasitoides*. *Helv. Chim. Acta* **2004**, 87, (4), 976-982.

112. Dirat, O.; Kouklovsky, C.; Langlois, Y., Oxazoline N-oxide-mediated [2+3] cycloadditions: Application to a total synthesis of the hypocholesterolemic agent 1233A. *J. Org. Chem.* **1998**, 63, (19), 6634-6642.

113. Chiang, Y. C. P.; Yang, S. S.; Heck, J. V.; Chabala, J. C.; Chang, M. N., Total synthesis of L-659,699, a novel inhibitor of cholesterol biosynthesis. *J. Org. Chem.* **1989**, 54, (24), 5708-5712.

114. Wohlrab, A.; Lamer, R.; van Nieuwenhze, M. S., Total synthesis of plusbacin A(3): A depsipeptide antibiotic active against vancomycin-resistant bacteria. *J. Am. Chem. Soc.* **2007**, 129, (14), 4175-4177.

115. Nihei, K.; Hashimoto, K.; Miyairi, K.; Okuno, T., Enantioselective synthesis of four isomers of 3-hydroxy-4-methyltetradecanoic acid, the constituent of antifungal cyclodepsipeptides W493 A and B. *Biosci., Biotechnol., Biochem.* **2005**, 69, (1), 231-234.

116. Yin, J.; Yang, X. B.; Chen, Z. X.; Zhang, Y. H., Total synthesis of (-)-tetrahydrolipstatin by the tandem Mukaiyama-aldol lactonization. *Chin. Chem. Lett.* **2005**, 16, (11), 1448-1450.

117. Pons, J. M.; Kocienski, P., A synthesis of (-)-tetrahydrolipstatin. *Tetrahedron Lett.* **1989**, 30, (14), 1833-1836.

118. Martin, O. R.; Zhou, W.; Wu, X.; Front-Deschamps, S.; Moutel, S.; Schindl, K.; Jeandet, P.; Zbaeren, C.; Bauer, J. A., Synthesis and immunobiological activity of an original series of acyclic lipid A mimics based on a pseudodipeptide backbone. *J. Med. Chem.* **2006**, 49, (20), 6000-6014.

119. Sarabia, F.; Chammaa, S., Synthetic studies on stevastelins - Total synthesis of stevastelins B and B3. *J. Org. Chem.* **2005**, 70, (20), 7846-7857.

120. Gupta, P.; Naidu, S. V.; Kumar, P., An efficient total synthesis of sulfobacin A. *Tetrahedron Lett.* **2004**, 45, (52), 9641-9643.

121. Irako, N.; Shioiri, T., Total synthesis of sulfobacin A (flavocristamide B). *Tetrahedron Lett.* **1998**, 39, (32), 5793-5796.

122. Labeeuw, O.; Phansavath, P.; Genet, J.-P., Total synthesis of sulfobacin A through dynamic kinetic resolution of a racemic [β]-keto-[α]-amino ester hydrochloride. *Tetrahedron: Asymmetry* **2004**, 15, (12), 1899-1908.

123. Kiho, T.; Nakayama, M.; Kogen, H., Total synthesis and NMR conformational study of signal peptidase II inhibitors, globomycin and SF-1902 A(5). *Tetrahedron* **2003**, 59, (10), 1685-1697.

124. Kiho, T.; Nakayama, M.; Yasuda, K.; Miyakoshi, S.; Inukai, M.; Kogen, H., Synthesis and antimicrobial activity of novel globomycin analogues. *Bioorg. Med. Chem. Lett.* **2003**, 13, (14), 2315-2318.

References

125. Kiho, T.; Nakayama, M.; Yasuda, K.; Miyakoshi, S.; Inukai, M.; Kogen, H., Structure-activity relationships of globomycin analogues as antibiotics. *Bioorg. Med. Chem.* **2004**, 12, (2), 337-361.
126. Rodriguez, M. J.; Belvo, M.; Morris, R.; Zeckner, D. J.; Current, W. L.; Sachs, R. K.; Zweifel, M. J., The synthesis of pseudomycin C via a novel acid promoted side-chain deacylation of pseudomycin A. *Bioorg. Med. Chem. Lett.* **2001**, 11, (2), 161-164.
127. Shioiri, T.; Terao, Y.; Irako, N.; Aoyama, T., Synthesis of topostins B567 and D654 (WB-3559D, flavolipin), DNA topoisomerase I inhibitors of bacterial origin. *Tetrahedron* **1998**, 54, (51), 15701-15710.
128. Katoh, O.; Sugai, T.; Ohta, H., Application of microbial enantiofacially selective hydrolysis in natural product synthesis. *Tetrahedron: Asymmetry* **1994**, 5, (10), 1935-1944.
129. Morgan, B.; Burk, M. Methods for making simvatatin and intermediates. WO/2005/040107, 2005.
130. Lee, S.; Lee, K. Method of preparing statins intermediates. WO/2004/096789, 2004.
131. Hiramoto, M.; Okada, K.; Nagai, S.; Kawamoto, H., Structure of viscosin, a peptide antibiotic - Syntheses of D- and L-3-hydroxyacyl-L-leucine hydrazides related to viscosin. *Chem. Pharm. Bull.* **1971**, 19, (7), 1308-1314.
132. Mori, K.; Otaka, K., Synthesis of sphingofungin D and its stereoisomer at C14. *Tetrahedron Lett.* **1994**, 35, (49), 9207-9210.
133. Vanmiddlesworth, F.; Dufresne, C.; Wincott, F. E.; Mosley, R. T.; Wilson, K. E., Determination of the relative and absolute stereochemistry of sphingofungin A, sphingofungin B, sphingofungin C, and sphingofungin D. *Tetrahedron Lett.* **1992**, 33, (3), 297-300.
134. Zlicar, M. Process for the synthesis of rosuvastatin calcium. WO2007/017117A1, 2007.
135. Wu, Y. K.; Sun, Y. P., Novel chemoselective tosylation of the alcoholic hydroxyl group of syn-α,β-disubstituted β-hydroxy carboxylic acids. *Chem. Commun.* **2005**, (14), 1906-1908.
136. Inoue, M.; Nakada, M., Structure elucidation and enantioselective total synthesis of the potent HMG-CoA reductase inhibitor FR901512 via catalytic asymmetric Nozaki-Hiyama reactions. *J. Am. Chem. Soc.* **2007**, 129, (14), 4164-5.
137. Kobayashi, S.; Matsumura, M.; Furuta, T.; Hayashi, T.; Iwamoto, S., The asymmetric synthesis of sphingofungin F and the determination of its stereochemistry. *Synlett* **1997**, (3), 301-303.
138. Crocker, P. J.; Miller, M. J., Oxidative free-radical cyclization as a method for annulating β-lactams: Syntheses of functionalized carbacephams. *J. Org. Chem.* **1995**, 60, (19), 6176-6179.
139. Paterson, I.; Hulme, A. N., Total synthesis of (-)-ebelactone A and (-)-ebelactone B1. *J. Org. Chem.* **1995**, 60, (11), 3288-3300.
140. Marotta, E.; Piombi, B.; Righi, P.; Rosini, G., N-bromosuccinimide-induced lactonization of bicyclo[3.2.0]hept-3-en-6-ones. *J. Org. Chem.* **1994**, 59, (24), 7526-7528.
141. Aebi, J. D.; Deyo, D. T.; Chong, Q. S.; Guillaume, D.; Dunlap, B.; Rich, D. H., Synthesis, conformation, and immunosuppressive activities of 3 analogs of cyclosporine A modified in the 1-position. *J. Med. Chem.* **1990**, 33, (3), 999-1009.
142. Colucci, W. J.; Tung, R. D.; Petri, J. A.; Rich, D. H., Synthesis of D-Lysine 8-cyclosporine A - Further characterization of Bop-Cl in the 2-7 hexapeptide fragment synthesis. *J. Org. Chem.* **1990**, 55, (9), 2895-2903.
143. Rich, D. H.; Sun, C. Q.; Guillaume, D.; Dunlap, B.; Evans, D. A.; Weber, A. E., Synthesis, biological activity, and conformational analysis of (2s,3r,4s)-mebmt1-cyclosporin, a novel 1-position epimer of cyclosporine A. *J. Med. Chem.* **1989**, 32, (8), 1982-1987.
144. Schreiber, S. L.; Anthony, N. J.; Dorsey, B. D.; Hawley, R. C., Is there a scaffolding domain within the structure of the immunosuppressive agent cyclosporin A (CsA)? Studies of the cyclophilin binding domain of CsA. *Tetrahedron Lett.* **1988**, 29, (50), 6577-6580.
145. Schmidt, U.; Siegel, W., Amino acids and peptides. Synthesis of (4R)-4-((E)-2-butenyl)-4,N-dimethyl-L-threonine (mebmt), the characteristic amino acid of cyclosporine. *Tetrahedron Lett.* **1987**, 28, (25), 2849-2852.
146. Lynch, J. E.; Volante, R. P.; Wattley, R. V.; Shinkai, I., Synthesis of an HMG-CoA reductase inhibitor - A diastereoselective aldol approach. *Tetrahedron Lett.* **1987**, 28, (13), 1385-1388.

147. Hirama, M.; Noda, T.; Ito, S., Convenient synthesis of (S)-citronellol of high optical purity. *J. Org. Chem.* **1985**, 50, (1), 127-129.

148. Ito, Y.; Ishida, K.; Okada, S.; Murakami, M., The absolute stereochemistry of anachelins, siderophores from the cyanobacterium *Anabaena cylindrica*. *Tetrahedron* **2004**, 60, (41), 9075-9080.

149. Keri, V.; Szabo, C.; Arvai, E.; Aronhime, J. Forms of pravastatin sodium. 09/736796, 2006.

150. Faveau, C.; Mondon, M.; Gesson, J. P.; Mahnke, T.; Gebhardt, S.; Koert, U., Synthetic studies on a phenyl-laulimalide analogue. *Tetrahedron Lett.* **2006**, 47, (47), 8305-8308.

151. Hattori, M.; Takai, H.; Kinoshita, M., Syntheses and condensation polymerizations of 3-hydroxybutyric acid-derivatives of pyrimidine-bases. *Macromol. Chem. Phys.* **1978**, 179, (4), 905-913.

152. Becker, G. O. H.; Berger, W.; Domschke, G., *Organikum organisch-chemisches Grundpraktikum*. 21 ed.; Wiley-VCH: Weinheim, 2001.

153. Kobayashi, T.; Hori, Y.; Kakimoto, M. A.; Imai, Y., Synthesis of biodegradable polyesters by polycondensation of methyl (R)-3-hydroxybutyrate and methyl (R)-3-hydroxyvalerate. *Macromol. Rapid Commun.* **1993**, 14, (12), 785-790.

154. Lengweiler, U. D.; Fritz, M. G.; Seebach, D., Synthesis of monodisperse linear and cyclic oligo[(R)-3-hydroxybutanoates] containing up to 128 monomeric units. *Helv. Chim. Acta* **1996**, 79, (3), 670-701.

155. Seebach, D.; Fritz, M. G., Detection, synthesis, structure, and function of oligo(3-hydroxyalkanoates): Contributions by synthetic organic chemists. *Int. J. Biol. Macromol.* **1999**, 25, (1-3), 217-236.

156. Maehara, A.; Taguchi, S.; Nishiyama, T.; Yamane, T.; Doi, Y., A repressor protein, PhaR, regulates polyhydroxyalkanoate (PHA) synthesis via its direct interaction with PHA. *J. Bacteriol.* **2002**, 184, (14), 3992-4002.

157. Curley, J. M.; Lenz, R. W.; Fuller, R. C., Sequential production of two different polyesters in the inclusion bodies of *Pseudomonas oleovorans*. *Int. J. Biol. Macromol.* **1996**, 19, (1), 29-34.

158. Lopez, N. I.; Floccari, M. E.; Steinbüchel, A.; Garcia, A. F.; Mendez, B. S., Effect of poly(3-hydroxybutyrate) (PHB) content on the starvation-survival of bacteria in natural waters. *FEMS Microbiol. Ecol.* **1995**, 16, (2), 95-101.

159. Ruiz, J. A.; Lopez, N. I.; Fernandez, R. O.; Mendez, B. S., Polyhydroxyalkanoate degradation is associated with nucleotide accumulation and enhances stress resistance and survival of *Pseudomonas oleovorans* in natural water microcosms. *Appl. Environ. Microbiol.* **2001**, 67, (1), 225-230.

160. Pham, T. H.; Webb, J. S.; Rehm, B. H. A., The role of polyhydroxyalkanoate biosynthesis by *Pseudomonas aeruginosa* in rhamnolipid and alginate production as well as stress tolerance and biofilm formation. *Microbiology* **2004**, 150, 3405-3413.

161. Matin, A.; Veldhuis, C.; Stegeman, V.; Veenhuis, M., Selective advantage of a *Spirillum sp.* in a carbon-limited environment - Accumulation of poly-β-hydroxybutyric acid and its role in starvation. *J. Gen. Microbiol.* **1979**, 112, 349-355.

162. Kessler, B.; Kraak, M. N.; Ren, Q.; Klinke, S.; Prieto, M.; Witholt, B., Enzymology and molecular genetics of mcl-PHA biosynthesis. In *Biochemical principles and mechanisms of biosynthesis and degradation of polymers*, Steinbüchel, A., Ed. Wiley-VCH: Weinheim, Germany, 1998; pp 48–56.

163. Findlay, R. H.; White, D. C., Polymeric β-hydroxyalkanoates from environmental-samples and *Bacillus megaterium*. *Appl. Environ. Microbiol.* **1983**, 45, (1), 71-78.

164. Kjelleberg, S.; Albertson, N.; Flardh, K.; Holmquist, L.; Jouperjaan, A.; Marouga, R.; Ostling, J.; Svenblad, B.; Weichart, D., How do nondifferentiating bacteria adapt to starvation. *A. v. Leeuwenhoek* **1993**, 63, (3-4), 333-341.

165. Henrysson, T.; Mccarty, P. L., Influence of the endogenous storage lipid poly-β-hydroxybutyrate on the reducing power availability during cometabolism of trichloroethylene and naphthalene by resting methanotrophic mixed cultures. *Appl. Environ. Microbiol.* **1993**, 59, (5), 1602-1606.

166. Senior, P. J.; Beech, G. A.; Ritchie, G. A. F.; Dawes, E. A., The role of oxygen limitation in the formation of poly-β-hydroxybutyrate during batch and continuous culture of *Azotobacter beijerinckii*. *Biochem. J.* **1972**, 128, 1193-1201.

167. Steinbüchel, A.; Füchtenbusch, B.; Gorenflo, V.; Hein, S.; Jossek, R.; Langenbach, S.; Rehm, B. H. A., Biosynthesis of polyesters in bacteria and recombinant organisms. *Polym. Degrad. Stab.* **1998**, 59, (1-3), 177-182.

References

168. Moskowitz, G. J.; Merrick, J. M., Metabolism of poly-β-hydroxybutyrate - Enzymatic synthesis of D-(-)-β-hydroxybutyryl coenzyme A by an enoyl hydrase from *Rhodospirillum rubrum. Biochemistry* **1969**, 8, (7), 2748-55.

169. Doi, Y., Polyhydroxyalkanoates. In *Handbook of Biodegradable Polymers*, Domb, A. J.; Kost, J.; Wiseman, D. M., Eds. Harwood Academic Publ GmbH: Chur, 1997; Vol. 7, pp 79-86.

170. Valentin, H. E.; Schönebaum, A.; Steinbüchel, A., Identification of 4-hydroxyvaleric acid as a constituent of biosynthetic polyhydroxyalkanoic acids from bacteria. *Appl. Microbiol. Biotechnol.* **1992**, 36, 507-514.

171. Schmack, G.; Gorenflo, V.; Steinbüchel, A., Biotechnological production and characterization of polyesters containing 4-hydroxyvaleric acid and medium-chain-length hydroxyalkanoic acids. *Macromolecules* **1998**, 31, (3), 644-649.

172. Byrom, D., Production of poly-β-hydroxybutyrate - poly-β-hydroxyvalerate copolymers. *FEMS Microbiol. Rev.* **1992**, 103, (2-4), 247-250.

173. Slater, S.; Mitsky, T. A.; Houmiel, K. L.; Hao, M.; Reiser, S. E.; Taylor, N. B.; Tran, M.; Valentin, H. E.; Rodriguez, D. J.; Stone, D. A.; Padgette, S. R.; Kishore, G.; Gruys, K. J., Metabolic engineering of *Arabidopsis* and *Brassica* for poly(3-hydroxybutyrate-co-3-hydroxyvalerate) copolymer production. *Nat. Biotechnol.* **1999**, 17, (10), 1011-1016.

174. Dawes, E. A.; Senior, P. J., The role and regulation of energy reserve polymers in microorganisms. In *Advances in Microbial Physiology*, Rose, A. H.; Tempest, D. W., Eds. Academic Press: London, 1973; Vol. 10, pp 135-266.

175. Schlegel, H. G.; Gottschalk, G., Poly-β-hydroxybuttersäure, ihre Verbreitung, Funktion und Biosynthese. *Angew. Chem.* **1962**, 74, 342-347.

176. Dawes, E. A., Polyhydroxybutyrate: An intriguing biopolymer. *Biosci. Rep.* **1988**, 8, (6), 537-547.

177. Jackson, F. A.; Dawes, E. A., Regulation of the tricarboxylic acid cycle and poly-ß-hydroxybutyrate metabolism in *Azotobacter beijerinckii* grown under nitrogen or oxygen limitation. *J. Gen. Microbiol.* **1976**, 97, 303-312.

178. de Smet, M. J.; Eggink, G.; Witholt, B.; Kingma, J.; Wynberg, H., Characterization of intracellular inclusions formed by *Pseudomonas oleovorans* during growth on octane. *J. Bacteriol.* **1983**, 154, (2), 870-878.

179. Huisman, G. W.; Deleeuw, O.; Eggink, G.; Witholt, B., Synthesis of poly-3-hydroxyalkanoates is a common feature of fluorescent pseudomonads. *Appl. Environ. Microbiol.* **1989**, 55, (8), 1949-1954.

180. Kunau, W. H.; Dommes, V.; Schulz, H., β-oxidation of fatty acids in mitochondria, peroxisomes, and bacteria: A century of continued progress. *Prog. Lipid Res.* **1995**, 34, (4), 267-342.

181. Fukui, T.; Shiomi, N.; Doi, Y., Expression and characterization of (*R*)-specific enoyl coenzyme A hydratase involved in polyhydroxyalkanoate biosynthesis by *Aeromonas caviae. J. Bacteriol.* **1998**, 180, (3), 667-673.

182. Tsuge, T.; Fukui, T.; Matsusaki, H.; Taguchi, S.; Kobayashi, G.; Ishizaki, A.; Doi, Y., Molecular cloning of two (*R*)-specific enoyl-CoA hydratase genes from *Pseudomonas aeruginosa* and their use for polyhydroxyalkanoate synthesis. *FEMS Microbiol. Lett.* **2000**, 184, (2), 193-198.

183. Langenbach, S.; Rehm, B. H. A.; Steinbüchel, A., Functional expression of the PHA synthase gene PhaC1 from *Pseudomonas aeruginosa* in *Escherichia coli* results in poly(3-hydroxyalkanoate) synthesis. *FEMS Microbiol. Lett.* **1997**, 150, (2), 303-309.

184. Qi, Q. S.; Rehm, B. H. A.; Steinbüchel, A., Synthesis of poly(3-hydroxyalkanoates) in *Escherichia coli* expressing the PHA synthase gene phaC2 from *Pseudomonas aeruginosa*: Comparison of PhaC1 and PhaC2. *FEMS Microbiol. Lett.* **1997**, 157, (1), 155-162.

185. Rehm, B. H. A.; Krüger, N.; Steinbüchel, A., A new metabolic link between fatty acid *de novo* synthesis and polyhydroxyalkanoic acid synthesis - The *phaG* gene from *Pseudomonas putida* KT2440 encodes a 3-hydroxyacyl-acyl carrier protein coenzyme A transferase. *J. Biol. Chem.* **1998**, 273, (37), 24044-24051.

186. Haywoold, G. W.; Anderson, A. J.; Ewing, D. F.; Dawes, E. A., Accumulation of a polyhydroxyalkanoate containing primarly 3-hydroxydecanoate from simple carbohydrate substrates by *Pseudomonas* sp. strain NCIMB 40135. *Appl. Environ. Microbiol.* **1990**, 56, (11), 3354-3359.

187. Timm, A.; Steinbüchel, A., Formation of polyesters consisting of medium-chain-length 3-hydroxyalkanoic acids from gluconate by *Pseudomonas aeruginosa* and other fluorescent pseudomonads. *Appl. Environ. Microbiol.* **1990**, 56, 3360-3367.

188. Huijberts, G. N. M.; Derijk, T. C.; Dewaard, P.; Eggink, G., ^{13}C nuclear-magnetic-resonance studies of *Pseudomonas putida* fatty acid metabolic routes involved in poly(3-hydroxyalkanoate) synthesis. *J. Bacteriol.* **1994**, 176, (6), 1661-1666.

189. Liu, S. J.; Steinbüchel, A., A novel genetically engineered pathway for synthesis of poly(hydroxyalkanoic acids) in *Escherichia coli*. *Appl. Environ. Microbiol.* **2000**, 66, (2), 739-743.

190. Ren, Q.; Sierro, N.; Kellerhals, M.; Kessler, B.; Witholt, B., Properties of engineered poly-3-hydroxyalkanoates produced in recombinant *Escherichia coli* strains. *Appl. Environ. Microbiol.* **2000**, 66, (4), 1311-1320.

191. Ren, Q.; Sierro, N.; Witholt, B.; Kessler, B., FabG, an NADPH-dependent 3-ketoacyl reductase of *Pseudomonas aeruginosa*, provides precursors for medium-chain-length poly-3-hydroxyalkanoate biosynthesis in *Escherichia coli*. *J. Bacteriol.* **2000**, 182, (10), 2978-2981.

192. Rehm, B. H. A.; Steinbüchel, A., Heterologous expression of the acyl-acyl carrier protein thioesterase gene from the plant *Umbellularia californica* mediates polyhydroxyalkanoate biosynthesis in recombinant *Escherichia coli*. *Appl. Microbiol. Biotechnol.* **2001**, 55, (2), 205-209.

193. Gao, D.; Maehara, A.; Yamane, T.; Ueda, S., Identification of the intracellular polyhydroxyalkanoate depolymerase gene of *Paracoccus denitrificans* and some properties of the gene product. *FEMS Microbiol. Lett.* **2001**, 196, (2), 159-164.

194. Li, Z. G.; Lin, H.; Ishii, N.; Chen, G. Q.; Inoue, Y., Study of enzymatic degradation of microbial copolyesters consisting of 3-hydroxybutyrate and medium-chain-length 3-hydroxyalkanoates. *Polym. Degrad. Stab.* **2007**, 92, (9), 1708-1714.

195. Kobayashi, T.; Shiraki, M.; Abe, T.; Sugiyama, A.; Saito, T., Purification and properties of an intracellular 3-hydroxybutyrate-oligomer hydrolase (PhaZ2) in *Ralstonia eutropha* H16 and its identification as a novel intracellular poly(3-hydroxybutyrate) depolymerase. *J. Bacteriol.* **2003**, 185, (12), 3485-3490.

196. Kobayashi, T.; Uchino, K.; Abe, T.; Yamazaki, Y.; Saito, T., Novel intracellular 3-hydroxybutyrate-oligomer hydrolase in *Wautersia eutropha* H16. *J. Bacteriol.* **2005**, 187, (15), 5129-5135.

197. Handrick, R.; Reinhardt, S.; Jendrossek, D., Mobilization of poly(3-hydroxybutyrate) in *Ralstonia eutropha*. *J. Bacteriol.* **2000**, 182, (20), 5916-5918.

198. Pötter, M.; Steinbüchel, A., Poly(3-hydroxybutyrate) granule-associated proteins: Impacts on poly(3-hydroxybutyrate) synthesis and degradation. *Biomacromolecules* **2005**, 6, (2), 552-560.

199. Gerngross, T. U.; Reilly, P.; Stubbe, J.; Sinskey, A. J.; Peoples, O. P., Immunocytochemical analysis of poly-β-hydroxybutyrate (PHB) synthase in *Alcaligenes eutrophus* H16: Localization of the synthase enzyme at the surface of the PHB granules. *J. Bacteriol.* **1993**, 175, 5289-5293.

200. Stuart, E. S.; Lenz, R. W.; Fuller, R. C., The ordered macromolecular surface of polyester inclusion bodies in *Pseudomonas oleovorans*. *Can. J. Microbiol.* **1995**, 41, (Suppl. 1), 84-93.

201. Liebergesell, M.; Sonomoto, K.; Madkour, M.; Mayer, F.; Steinbüchel, A., Purification and characterization of the poly(hydroxyalkanoic acid) synthase from *Chromatium vinosum* and localization of the enzyme at the surface of poly(hydroxyalkanoic acid) granules. *Eur. J. Biochem.* **1994**, 226, (1), 71-80.

202. Rehm, B. H. A., Polyester synthases: Natural catalysts for plastics. *Biochem. J.* **2003**, 376, 15-33.

203. Liebergesell, M.; Mayer, F.; Steinbüchel, A., Analysis of polyhydroxyalkanoic acid-biosynthesis genes of anoxygenic phototrophic bacteria reveals synthesis of a polyester exhibiting an unusual composition. *Appl. Microbiol. Biotechnol.* **1993**, 40, 292-300.

204. Doi, Y.; Kitamura, S.; Abe, H., Microbial synthesis and characterization of poly(3-hydroxybutyrate-co-3-hydroxyhexanoate). *Macromolecules* **1995**, 28, (14), 4822-4828.

205. Dennis, D.; McCoy, M.; Stangl, A.; Valentin, H. E.; Wu, Z., Formation of poly(3-hydroxybutyrate-co-3-hydroxyhexanoate) by PHA synthase from *Ralstonia eutropha*. *J. Biotechnol.* **1998**, 64, (2-3), 177-186.

206. Antonio, R. V.; Steinbüchel, A.; Rehm, B. H. A., Analysis of *in vivo* substrate specificity of the PHA synthase from *Ralstonia eutropha*: Formation of novel copolyesters in recombinant *Escherichia coli*. *FEMS Microbiol. Lett.* **2000**, 182, (1), 111-117.

References

207. Rehm, B. H. A.; Steinbüchel, A., Biochemical and genetic analysis of PHA synthases and other proteins required for PHA synthesis. *Int. J. Biol. Macromol.* **1999**, 25, (1-3), 3-19.

208. Hoppensack, A.; Rehm, B. H. A.; Steinbüchel, A., Analysis of 4-phosphopantetheinylation of polyhydroxybutyrate synthase from *Ralstonia eutropha*: Generation of β-alanine auxotrophic Tn5 mutants and cloning of the *panD* gene region. *J. Bacteriol.* **1999**, 181, (5), 1429-1435.

209. Müh, U.; Sinskey, A. J.; Kirby, D. P.; Lane, W. S.; Stubbe, J. A., PHA synthase from *Chromatium vinosum*: Cysteine 149 is involved in covalent catalysis. *Biochemistry* **1999**, 38, (2), 826-837.

210. Timm, A.; Byrom, D.; Steinbüchel, A., Formation of blends of various poly(3-hydroxyalkanoic acids) by a recombinant strain of *Pseudomonas oleovorans*. *Appl. Microbiol. Biotechnol.* **1990**, 33, 296-301.

211. Jossek, R.; Steinbüchel, A., In vitro synthesis of poly(3-hydroxybutyric acid) by using an enzymatic coenzyme A recycling system. *FEMS Microbiol. Lett.* **1998**, 168, (2), 319-324.

212. Lemoigne, Biological chemistry. The origin of β-oxybutyric acid obtained using the microbial process. *C. R. l'Academie. Sci.* **1925**, 180, 1539-1541.

213. Macrae, R. M.; Wilkinson, J., Poly-ß-hydroxybutyrate metabolism in washed suspension of *Bacillus cereus* and *Bacillus megaterium*. *J. Gen. Microbiol.* **1958**, 19, 210-222.

214. Hippe, H., Abbau und Wiederverwertung von Poly-β-hydroxybuttersäure durch *Hydrogenomonas* H16. *Arch. Microbiol.* **1967**, 56, 248-277.

215. Choi, M. H.; Yoon, S. C.; Lenz, R. W., Production of poly(3-hydroxybutyric acid-co-4-hydroxybutyric acid) and poly(4-hydroxybutyric acid) without subsequent degradation by *Hydrogenophaga pseudoflava*. *Appl. Environ. Microbiol.* **1999**, 65, (4), 1570-1577.

216. Foster, L. J. R.; Lenz, R. W.; Fuller, R. C., Quantitative determination of intracellular depolymerase activity in *Pseudomonas oleovorans* inclusions containing poly-3-hydroxyalkanoates with long alkyl substituents. *FEMS Microbiol. Lett.* **1994**, 118, (3), 279-282.

217. Foster, L. J. R.; Stuart, E. S.; Tehrani, A.; Lenz, R. W.; Fuller, R. C., Intracellular depolymerase and polyhydroxyoctanoate granule integrity in *Pseudomonas oleovorans*. *Int. J. Biol. Macromol.* **1996**, 19, (3), 177-183.

218. Stuart, E. S.; Foster, L. J. R.; Lenz, R. W.; Fuller, R. C., Intracellular depolymerase functionality and location in *Pseudomonas oleovorans* inclusions containing polyhydroxyoctanoate. *Int. J. Biol. Macromol.* **1996**, 19, (3), 171-176.

219. Handrick, R.; Reinhardt, S.; Kimmig, P.; Jendrossek, D., The "intracellular" poly(3-hydroxybutyrate) (PHB) depolymerase of *Rhodospirillum rubrum* is a periplasm-located protein with specificity for native PHB and with structural similarity to extracellular PHB depolymerases. *J. Bacteriol.* **2004**, 186, (21), 7243-7253.

220. Saegusa, H.; Shiraki, M.; Kanai, C.; Saito, T., Cloning of an intracellular poly[D-(-)-3-hydroxybutyrate] depolymerase gene from *Ralstonia eutropha* H16 and characterization of the gene product. *J. Bacteriol.* **2001**, 183, (1), 94-100.

221. Saito, T.; Takizawa, K.; Saegusa, H., Intracellular poly(3-hydroxybutyrate) depolymerase in *Alcaligenes eutrophus*. *Can. J. Microbiol.* **1995**, 41, (Suppl. 1), 187-191.

222. Tseng, C. L.; Chen, H. J.; Shaw, G. C., Identification and characterization of the *Bacillus thuringiensis phaZ* gene, encoding new intracellular poly-3-hydroxybutyrate depolymerase. **2006**, 188, (21), 7592-7599.

223. York, G. M.; Lupberger, J.; Tian, J. M.; Lawrence, A. G.; Stubbe, J.; Sinskey, A. J., *Ralstonia eutropha* H16 encodes two and possibly three intracellular poly[D-(-)-3-hydroxybutyrate] depolymerase genes. *J. Bacteriol.* **2003**, 185, (13), 3788-3794.

224. Sandoval, A.; Arias-Barrau, E.; Bermejo, F.; Canedo, L.; Naharro, G.; Olivera, E.; Luengo, J., Production of 3-hydroxy-n-phenylalkanoic acids by a genetically engineered strain of *Pseudomonas putida*. *Appl. Microbiol. Biotechnol.* **2005**, 67, (1), 97-105.

225. Jendrossek, D.; Schirmer, A.; Schlegel, H. G., Biodegradation of polyhydroxyalkanoic acids. *Appl. Microbiol. Biotechnol.* **1996**, 46, (5-6), 451-463.

226. Amor, S. R.; Rayment, T.; Sanders, J. K. M., Poly(hydroxybutyrate) *in vivo*: NMR and X-ray characterization of the elastomeric state. *Macromolecules* **1991**, 24, 4583-4588.

227. de Koning, G. J. M.; Lemstra, P. J., The amorphous state of bacterial poly[(R)-3-hydroxyalkanoate] in vivo. Polymer 1992, 33, (15), 3292-3294.

228. Numata, K.; Kikkawa, Y.; Tsuge, T.; Iwata, T.; Doi, Y.; Abe, H., Adsorption of biopolyester depolymerase on silicon wafer and poly[(R)-3-hydroxybutyric acid] single crystal revealed by real-time AFM. Macromol. Biosci. 2006, 6, (1), 41-50.

229. Schirmer, A.; Jendrossek, D., Molecular characterization of the extracellular poly(3-hydroxyoctanoic acid) [P(3HO)] depolymerase gene of Pseudomonas fluorescens GK13 and of its gene product. J. Bacteriol. 1994, 176, (22), 7065-7073.

230. Yamashita, K.; Aoyagi, Y.; Abe, H.; Doi, Y., Analysis of adsorption function of polyhydroxybutyrate depolymerase from Alcaligenes faecalis T1 by using a quartz crystal microbalance. Biomacromolecules 2001, 2, (1), 25-28.

231. Jendrossek, D.; Knoke, I.; Habibian, R. B.; Steinbüchel, A.; Schlegel, H. G., Degradation of poly(3-hydroxybutyrate), PHB, by bacteria and purification of a novel PHB depolymerase from Comamonas sp. J. Environm. Polym. Degrad. 1993, 1, 53-61.

232. Dirusso, C. C., Primary sequence of the Escherichia coli fadBA operon, encoding the fatty acid-oxidizing multienzyme complex, indicates a high degree of homology to eukaryotic enzymes. J. Bacteriol. 1990, 172, (11), 6459-6468.

233. Nunn, W. D., A molecular view of fatty acid catabolism in Escherichia coli. Microbiol. Rev. 1986, 50, (2), 179-192.

234. Starai, V. J.; Escalante-Semerena, J. C., Acetyl-coenzyme A synthetase (AMP forming). Cell. Mol. Life Sci. 2004, 61, (16), 2020-2030.

235. Dewet, J. R.; Wood, K. V.; Deluca, M.; Helinski, D. R.; Subramani, S., Firefly luciferase gene - Structure and expression in mammalian cells. Mol. Cell. Biol. 1987, 7, (2), 725-737.

236. Duronio, R. J.; Knoll, L. J.; Gordon, J. I., Isolation of a Saccharomyces cerevisiae long-chain fatty acyl-CoA synthetase gene (faa1) and assessment of its role in protein N-myristoylation. J. Cell Biol. 1992, 117, (3), 515-529.

237. Suzuki, H.; Kawarabayasi, Y.; Kondo, J.; Abe, T.; Nishikawa, K.; Kimura, S.; Hashimoto, T.; Yamamoto, T., Structure and regulation of rat long-chain acyl-CoA synthetase. J. Biol. Chem. 1990, 265, (15), 8681-8685.

238. Saraste, M.; Sibbald, P. R.; Wittinghofer, A., The P-loop - A common motif in ATP-binding and GTP-binding proteins. Trends Biochem. Sci. 1990, 15, (11), 430-434.

239. Black, P. N.; DiRusso, C. C.; Metzger, A. K.; Heimert, T. L., Cloning, sequencing, and expression of the fadD gene of Escherichia coli encoding acyl-coenzyme A synthetase. J. Biol. Chem. 1992, 267, (35), 25513-25520.

240. Overath, P.; Pauli, G.; Schairer, H. U., Fatty acid degradation in Escherichia coli - An inducible acyl-CoA synthetase mapping of old-mutations and isolation of regulatory mutants. Eur. J. Biochem. 1969, 7, (4), 559-574.

241. Black, P. N., Characterization of FadL-specific fatty acid binding in Escherichia coli. Biochim. Biophys. Acta 1990, 1046, (1), 97-105.

242. Nelson, K. E.; Weinel, C.; Paulsen, I. T.; Dodson, R. J.; Hilbert, H.; dos Santos, V.; Fouts, D. E.; Gill, S. R.; Pop, M.; Holmes, M.; Brinkac, L.; Beanan, M.; DeBoy, R. T.; Daugherty, S.; Kolonay, J.; Madupu, R.; Nelson, W.; White, O.; Peterson, J.; Khouri, H.; Hance, I.; Lee, P. C.; Holtzapple, E.; Scanlan, D.; Tran, K.; Moazzez, A.; Utterback, T.; Rizzo, M.; Lee, K.; Kosack, D.; Moestl, D.; Wedler, H.; Lauber, J.; Stjepandic, D.; Hoheisel, J.; Straetz, M.; Heim, S.; Kiewitz, C.; Eisen, J.; Timmis, K. N.; Dusterhoft, A.; Tummler, B.; Fraser, C. M., Complete genome sequence and comparative analysis of the metabolically versatile Pseudomonas putida KT2440. Environ. Microbiol. 2002, 4, (12), 799-808.

243. Riley, M.; Abe, T.; Arnaud, M. B.; Berlyn, M. K. B.; Blattner, F. R.; Chaudhuri, R. R.; Glasner, J. D.; Horiuchi, T.; Keseler, I. M.; Kosuge, T.; Mori, H.; Perna, N. T.; Plunkett, G.; Rudd, K. E.; Serres, M. H.; Thomas, G. H.; Thomson, N. R.; Wishart, D.; Wanner, B. L., Escherichia coli K12: A cooperatively developed annotation snapshot - 2005. Nucleic Acids Res. 2006, 34, (1), 1-9.

244. Fernandez-Valverde, M.; Reglero, A.; Martinezblanco, H.; Luengo, J. M., Purification of Pseudomonas putida acyl-coenzyme A ligase active with a range of aliphatic and aromatic substrates. Appl. Environ. Microbiol. 1993, 59, (4), 1149-1154.

References

245. Arias-Barrau, E.; Olivera, E. R.; Sandoval, A.; Naharro, G.; Luengo, J. M., Acetyl-CoA synthetase from *Pseudomonas putida* U is the only acyl-CoA activating enzyme induced by acetate in this bacterium. *FEMS Microbiol. Lett.* **2006**, 260, (1), 36-46.

246. Martinezblanco, H.; Reglero, A.; Rodriguezaparicio, L. B.; Luengo, J. M., Purification and biochemical characterization of phenylacetyl-CoA ligase from *Pseudomonas putida* - A specific enzyme for the catabolism of phenylacetic acid. *J. Biol. Chem.* **1990**, 265, (12), 7084-7090.

247. Hoffmann, N.; Amara, A. A.; Beermann, B. B.; Qi, Q. S.; Hinz, H. J.; Rehm, B. H. A., Biochemical characterization of the *Pseudomonas putida* 3-hydroxyacyl ACP : CoA transacylase, which diverts intermediates of fatty acid *de novo* biosynthesis. *J. Biol. Chem.* **2002**, 277, (45), 42926-42936.

248. Olivera, E. R.; Carnicero, D.; Garcia, B.; Minambres, B.; Moreno, M. A.; Canedo, L.; DiRusso, C. C.; Naharro, G.; Luengo, J. M., Two different pathways are involved in the β-oxidation of *n*-alkanoic and *n*-phenylalkanoic acids in *Pseudomonas putida* U: Genetic studies and biotechnological applications. *Mol. Microbiol.* **2001**, 39, (4), 863-874.

249. Lee, S. Y.; Lee, Y. H., Metabolic engineering of *Escherichia coli* for production of enantiomerically pure (R)-(-)-hydroxycarboxylic acids. *Appl. Environ. Microbiol.* **2003**, 69, (6), 3421-3426.

250. Furrer, P.; Hany, R.; Rentsch, D.; Grubelnik, A.; Ruth, K.; Panke, S.; Zinn, M., Quantitative analysis of bacterial medium-chain-length poly([R]-3-hydroxyalkanoates) by gas chromatography. *J. Chromatogr. A* **2007**, 1143, (1-2), 199-206.

251. Breeuwer, P.; Abee, T., Assessment of the membrane potential, intracellular pH and respiration of bacteria employing fluorescence techniques. *Mol. Mirob. Ecol. Manual* **2004**, 2, (8.01), 1563-1580.

252. Witholt, B.; Kessler, B., Perspectives of medium-chain-length poly(hydroxyalkanoates), a versatile set of bacterial bioplastics. *Curr. Opin. Biotechnol.* **1999**, 10, (3), 279-285.

253. Blaser, H. U., Enantioselective catalysis in fine chemicals production. *Chem. Commun.* **2003**, (3), 293-296.

254. Huisman, G. W.; Wonink, E.; de Koning, G. J. M.; Preusting, H.; Witholt, B., Synthesis of poly (3-hydroxyalkanoates) by mutant and recombinant *Pseudomonas* strains. *Appl. Environ. Microbiol.* **1992**, 38, 1-5.

255. Riis, V.; Mai, W., Gas chromatographic determination of poly-β-hydroxybutyric acid in microbial biomass after hydrochloric acid propanolysis. *J. Chromatogr. A* **1988**, 445, 285-289.

256. Lageveen, R. G.; Huisman, G. W.; Preusting, H.; Ketelaar, P.; Eggink, G.; Witholt, B., Formation of polyesters by *Pseudomonas oleovorans*: Effect of substrates on formation and composition of poly-(R)-3-hydroxyalkanoates and poly-(R)-3-hydroxyalkenoates. *Appl. Environ. Microbiol.* **1988**, 54, 2924-2932.

257. Hazenberg, W.; Witholt, B., Efficient production of medium-chain-length poly(3-hydroxyalkanoates) from octane by *Pseudomonas oleovorans*: Economic considerations. *Appl. Microbiol. Biotechnol.* **1997**, 48, 588-596.

258. Hayward, A. C.; Forsyth, W. G. C.; Roberts, J. B., Synthesis and breakdown of poly-β-hydroxybutyric acid by bacteria. *J. Gen. Microbiol.* **1959**, 20, (3), 510-518.

259. Solaiman, D. K. Y.; Ashby, R. D.; Foglia, T. A., Effect of inactivation of poly(hydroxyalkanoates) depolymerase gene on the properties of poly(hydroxyalkanoates) in *Pseudomonas resinovorans*. *Appl. Microbiol. Biotechnol.* **2003**, 62, (5-6), 536-543.

260. Seidel, W.; Seebach, D., Grahamimycin a1 synthesis and determination of configuration and chirality. *Tetrahedron Lett.* **1982**, 23, (2), 159-162.

261. Seuring, B.; Seebach, D., Syntheses and determinations of the absolute configurations of norpyrenophorin, pyrenophorin, and vermiculine. *Liebigs Ann. Chem.* **1978**, (12), 2044-2073.

262. Sutter, M. A.; Seebach, D., Synthesis of (2E,4E,6S,7R,10E,12E,14S,15R)-6,7,14,15-tetramethyl-8,16-dioxa-2,4,10,12-cyclohexadecatetraene-1,9-dione- A model system for elaiophylin. *Liebigs Ann. Chem.* **1983**, (6), 939-949.

263. Burke, T. R.; Knight, M.; Chandrasekhar, B., Solid-phase synthesis of viscosin, a cyclic depsipeptide with antibacterial and antiviral properties. *Tetrahedron Lett.* **1989**, 30, (5), 519-522.

264. Ren, Q.; Grubelnik, A.; Hörler, M.; Ruth, K.; Hartmann, R.; Felber, H.; Zinn, M., Bacterial poly(hydroxyalkanoates) as a source of chiral hydroxyalkanoic acids. *Biomacromolecules* **2005**, 6, (4), 2290-2298.

265. Hartmann, R.; Hany, R.; Pletscher, E.; Ritter, A.; Witholt, B.; Zinn, M., Tailor-made olefinic medium-chain-length poly[(R)-3-hydroxyalkanoates] by *Pseudomonas putida* GPo1: Batch versus chemostat production. *Biotechnol. Bioeng.* **2006**, 93, (4), 737-746.

266. Floriano, B.; Ruiz-Barba, J. L.; Jimenez-Diaz, R., Purification and genetic characterization of enterocin I from *Enterococcus faecium* 6T1a, a novel antilisterial plasmid-encoded bacteriocin which does not belong to the pediocin family of bacteriocins. *Appl. Environ. Microbiol.* **1998**, 64, (12), 4883-4890.

267. Igarashi, M.; Tsuzuki, T.; Kambe, T.; Miyazawa, T., Recommended methods of fatty acid methylester preparation for conjugated dienes and trienes in food and biological samples. *J. Nutr. Sci. Vitaminol.* **2004**, 50, (2), 121-128.

268. Sheldon, R. A., Biocatalytic versus chemical synthesis of enantiomerically pure compounds. *Chimia* **1996**, 50, (9), 418-419.

269. Ohashi, T.; Hasegawa, J., D(−)-β-Hydroxycarboxylic acids as raw materials for captopril and β-lactams. In *Chirality in Industry*, Collins, A. N.; Sheldrake, G. N.; Crosby, J., Eds. Zeneca Specialities: Manchester, UK, 1992; pp 269-278.

270. Ruth, K.; Grubelnik, A.; Hartmann, R.; Egli, T.; Zinn, M.; Ren, Q., Efficient production of (R)-3-hydroxycarboxylic acids by biotechnological conversion of polyhydroxyalkanoates and their purification. *Biomacromolecules* **2007**, 8, (1), 279-286.

271. Doi, Y., *Poly(3-hydroxyalkanoates) metabolism*. VCH: Weinheim, Germany, 1990; p 63-88.

272. Kadouri, D.; Jurkevitch, E.; Okon, Y.; Castro-Sowinski, S., Ecological and agricultural significance of bacterial polyhydroxyalkanoates. *Crit. Rev. Microbiol.* **2005**, 31, (2), 55-67.

273. Senior, P. J.; Dawes, E. A., Poly-β-hydroxybutyrate biosynthesis and regulation of glucose metabolim in *Azotobacter beijerinckii*. *Biochem. J.* **1971**, 125, (1), 55-66.

274. Hirsch, M.; Elliott, T., Role of ppGpp in *rpoS* stationary-phase regulation in *Escherichia coli*. *J. Bacteriol.* **2002**, 184, (18), 5077-5087.

275. Lange, R.; Fischer, D.; Hengge-Aronis, R., Identification of transcriptional start sites and the role of ppGpp in the expression of RpoS, the structural gene for the σ^S subunit of RNA-polymerase in *Escherichia coli*. *J. Bacteriol.* **1995**, 177, (16), 4676-4680.

276. Ihssen, J.; Egli, T., Specific growth rate and not cell density controls the general stress response in *Escherichia coli*. *Microbiology* **2004**, 150, (6), 1637-1648.

277. Hengge-Aronis, R., Signal transduction and regulatory mechanisms involved in control of the σ^S (RpoS) subunit of RNA polymerase. *Microb. Mol. Biol. Rev.* **2002**, 66, (3), 373-395.

278. Peralta-Gil, M.; Segura, D.; Guzman, J.; Servin-Gonzalez, L.; Espin, G., Expression of the *Azotobacter vinelandii* poly-β-hydroxybutyrate biosynthetic *phbBAC* operon is driven by two overlapping promoters and is dependent on the transcriptional activator PhbR. *J. Bacteriol.* **2002**, 184, (20), 5672-5677.

279. Rehm, B. H. A., Genetics and biochemistry of polyhydroxyalkanoate granule self-assembly: The key role of polyester synthases. *Biotechnol. Lett.* **2006**, 28, (4), 207-213.

280. Handrick, R.; Reinhardt, S.; Schultheiss, D.; Reichart, T.; Schuler, D.; Jendrossek, V.; Jendrossek, D., Unraveling the function of the *Rhodospirillum rubrum* activator of polyhydroxybutyrate (PHB) degradation: The activator is a PHB-granule-bound protein (phasin). *J. Bacteriol.* **2004**, 186, (8), 2466-2475.

281. Jurasek, L.; Marchessault, R. H., The role of phasins in the morphogenesis of poly(3-hydroxybutyrate) granules. *Biomacromolecules* **2002**, 3, (2), 256-261.

282. Pötter, M.; Müller, H.; Reinecke, F.; Wieczorek, R.; Fricke, F.; Bowien, B.; Friedrich, B.; Steinbüchel, A., The complex structure of polyhydroxybutyrate (PHB) granules: Four orthologous and paralogous phasins occur in *Ralstonia eutropha*. *Microbiology* **2004**, 150, 2301-2311.

283. Hanahan, D., Studies on transformation of *Escherichia coli* with plasmids. *J. Mol. Biol.* **1983**, 166, (4), 557-580.

284. Bachmann, B., Derivations and genotypes of some mutant derivatives of *Escherichia coli* K12. In *Escherichia coli and Salmonella typhimurium: Cellular and molecular biology*, Neidhardt, F. C.; Ingraham, J. L.; Low, K. B.; Magasanik, B.;

References

Schaechter, M.; Umbarger, H. E., Eds. American Society for Microbiology: Washington D.C., 1987; Vol. 2, pp 1190-1219.

285. Rhie, H. G.; Dennis, D., Role of *fadR* and *atoC(Con)* mutations in poly(3-hydroxybutyrate-co-3-hydroxyvalerate) synthesis in recombinant PHA(+) *Escherichia coli*. *Appl. Environ. Microbiol.* **1995**, 61, (7), 2487-2492.

286. Schwartz, R. D.; Mccoy, C. J., *Pseudomonas oleovorans* hydroxylation-epoxidation system - Additional strain improvements. *Appl. Microbiol.* **1973**, 26, (2), 217-218.

287. West, S. E. H.; Schweizer, H. P.; Dall, C.; Sample, A. K.; Runyenjanecky, L. J., Construction of improved *Escherichia Pseudomonas* shuttle vectors derived from pUC18/19 and sequence of the region required for their replication in *Pseudomonas aeruginosa*. *Gene* **1994**, 148, (1), 81-86.

288. Sambrook, J.; Russel, D. W., *Molecular Cloning - A laboratory manual*. 3 ed.; Cold Spring Harbor Laboratory Press: New York, 2001; Vol. 1-3.

289. de Roo, G.; Ren, Q.; Witholt, B.; Kessler, B., Development of an improved in vitro activity assay for medium-chain-length PHA polymerases based on Coenzyme A release measurements. *J. Microbiol. Meth.* **2000**, 41, (1), 1-8.

290. Laemmli, U. K., Cleavage of structural proteins during assembly of head of bacteriophage T4. *Nature* **1970**, 227, (5259), 680-685.

291. Kraak, M. N.; Kessler, B.; Witholt, B., In vitro activities of granule-bound poly[(R)-3-hydroxyalkanoate] polymerase C1 of *Pseudomonas oleovorans* - Development of an activity test for medium-chain-length-poly(3-hydroxyalkanoate) polymerases. *Eur. J. Biochem.* **1997**, 250, (2), 432-439.

292. Spiekermann, P.; Rehm, B. H. A.; Kalscheuer, R.; Baumeister, D.; Steinbüchel, A., A sensitive, viable-colony staining method using Nile red for direct screening of bacteria that accumulate polyhydroxyalkanoic acids and other lipid storage compounds. *Arch. Microbiol.* **1999**, 171, (2), 73-80.

293. Barnard, G. C.; McCool, J. D.; Wood, D. W.; Gerngross, T. U., Integrated recombinant protein expression and purification platform based on *Ralstonia eutropha*. *Appl. Environ. Microbiol.* **2005**, 71, (10), 5735-5742.

294. Minambres, B.; Martinez-Blanco, H.; Olivera, E. R.; Garcia, B.; Diez, B.; Barredo, J. L.; Moreno, M. A.; Schleissner, C.; Salto, F.; Luengo, J. M., Molecular cloning and expression in different microbes of the DNA encoding *Pseudomonas putida* U phenylacetyl-CoA ligase - Use of this gene to improve the rate of benzylpenicillin biosynthesis in *Penicillium chrysogenum*. *J. Biol. Chem.* **1996**, 271, (52), 33531-33538.

295. Hayashi, K.; Morooka, N.; Yamamoto, Y.; Fujita, K.; Isono, K.; Choi, S.; Ohtsubo, E.; Baba, T.; Wanner, B. L.; Mori, H. H.; Oriuchi, T., Highly accurate genome sequences of *Escherichia coli* K12 strains MG1655 and W3110. *Mol. Syst. Biol.* **2006**, 2.

296. Jendrossek, D.; Selchow, O.; Hoppert, M., Poly(3-hydroxybutyrate) granules at the early stages of formation are localized close to the cytoplasmic membrane in *Caryophanon latum*. *Appl. Environ. Microbiol.* **2007**, 73, (2), 586-593.

297. Michinaka, Y.; Shimauchi, T.; Aki, T.; Nakajima, T.; Kawamoto, S.; Shigeta, S.; Suzuki, O.; Ono, K., Extracellular secretion of free fatty acids by disruption of a fatty acyl-CoA synthetase gene in *Saccharomyces cerevisiae*. *J. Biosci. Bioeng.* **2003**, 95, (5), 435-440.

298. Huijberts, G. N. M.; Eggink, G.; Dewaard, P.; Huisman, G. W.; Witholt, B., *Pseudomonas putida* KT2442 cultivated on glucose accumulates poly(3-hydroxyalkanoates) consisting of saturated and unsaturated monomers. *Appl. Environ. Microbiol.* **1992**, 58, (2), 536-544.

299. Saegusa, H.; Shiraki, M.; Saito, T., Cloning of an intracellular D(-)-3-hydroxybutyrate-oligomer hydrolase gene from *Ralstonia eutropha* H16 and identification of the active site serine residue by site-directed mutagenesis. *J. Biosci. Bioeng.* **2002**, 94, (2), 106-112.

300. Doi, Y., Microbial synthesis, physical properties, and biodegradability of polyhydroxyalkanoates. *Macromol. Symposia* **1995**, 98, 585-599.

301. Brandl, H.; Gross, R. A.; Lenz, R. W.; Fuller, R. C., *Pseudomonas oleovorans* as a source of poly(β-hydroxyalkanoates) for potential applications as biodegradable polyesters. *Appl. Environ. Microbiol.* **1988**, 54, (8), 1977-1982.

References

302. Kim, T. K.; Jung, Y. M.; Vo, M. T.; Shioya, S.; Lee, Y. H., Metabolic engineering and characterization of *phaC1* and *phaC2* genes from *Pseudomonas putida* KCTC1639 for overproduction of medium-chain-length polyhydroxyalkanoate. *Biotechnol. Progr.* **2006**, 22, (6), 1541-1546.

303. de Roo, G. Physiological basis of polyhydroxyalkanoate metabolism in *Pseudomonas putida*. PhD thesis, ETH Zürich, Zürich, 2002.

304. Liu, W. K.; Chen, G. Q., Production and characterization of medium-chain-length polyhydroxyalkanoate with high 3-hydroxytetradecanoate monomer content by *fadB* and *fadA* knockout mutant of *Pseudomonas putida* KT2442. *Appl. Microbiol. Biotechnol.* **2007**, 76, (5), 1153-1159.

305. Vo, M. T.; Lee, K. W.; Kim, T. K.; Lee, Y. H., Utilization of *fadA* knockout mutant *Pseudomonas putida* for overproduction of medium chain-length-polyhydroxyalkanoate. *Biotechnol. Lett.* **2007**, 29, (12), 1915-1920.

306. Simon, R.; Priefer, U.; Puhler, A., A broad host range mobilization system for *in vivo* genetic-engineering-transposon mutagenesis in gram-negative bacteria. *Biotechnology* **1983**, 1, (9), 784-791.

307. Metcalfe, L. D.; Schmitz, A. A., Rapid preparation of fatty acid esters for gas chromatographic analysis. *Anal. Chem.* **1961**, 33, (3), 363-4.

308. Sabirova, J. S.; Ferrer, M.; Lunsdorf, H.; Wray, V.; Kalscheuer, R.; Steinbuchel, A.; Timmis, K. N.; Golyshin, P. N., Mutation in a "*tesB*-like" hydroxyacyl-coenzyme A-specific thioesterase gene causes hyperproduction of extracellular polyhydroxyalkanoates by *Alcanivorax borkumensis* SK2. *J. Bacteriol.* **2006**, 188, (24), 8452-8459.

309. Berg, J. M.; Tymoczko, J. L.; Lubert Stryer, L.; Clarke, N. D., *Biochemistry*. 3 ed.; W.H. Freeman and Company: New York, 1995.

310. Sandoval, A.; Arias-Barrau, E.; Arcos, M.; Naharro, G.; Olivera, E. R.; Luengo, J. M., Genetic and ultrastructural analysis of different mutants of *Pseudomonas putida* affected in the poly-3-hydroxy-*n*-alkanoate gene cluster. *Environ. Microbiol.* **2007**, 9, (3), 737-751.

311. Ruiz, J. A.; Lopez, N. I.; Mendez, B. S., *rpoS* gene expression in carbon-starved cultures of the polyhydroxyalkanoate-accumulating species *Pseudomonas oleovorans*. *Curr. Microbiol.* **2004**, 48, (6), 396-400.

312. Zheng, Z.; Li, M.; Xue, X. J.; Tian, H. L.; Li, Z.; Chen, G. Q., Mutation on N-terminus of polyhydroxybutyrate synthase of *Ralstonia eutropha* enhanced PHB accumulation. *Appl. Microbiol. Biotechnol.* **2006**, 72, (5), 896-905.

313. Guseo, R.; Dalla Valle, A.; Guidolin, M., World oil depletion models: Price effects compared with strategic or technological interventions. *Technol. Forecast Soc.* **2007**, 74, (4), 452-469.

314. Zinn, M.; Hany, R., Tailored material properties of polyhydroxyalkanoates through biosynthesis and chemical modification. *Adv. Eng. Mater.* **2005**, 7, (5), 408-411.

315. Sun, Z. Y.; Ramsay, J. A.; Guay, M.; Ramsay, B. A., Fermentation process development for the production of medium-chain-length poly-3-hydroxyalkanoates. *Appl. Microbiol. Biotechnol.* **2007**, 75, (3), 475-485.

316. Kircher, M., White Biotechnology: Ready to partner and invest in. *Biotechnol. J.* **2006**, 1, (7), 787-794.

317. Rijk, T.; van de Meer, P.; Eggink, G.; Weusthuis, R., Methods for analysis of poly(3-hydroxyalkanoate) composition. In *Biopolymers*, Doi, Y.; Steinbüchel, A., Eds. Wiley-VCH Verlag GmbH: Weinheim, 2002; Vol. 3b, pp 1-21.

Appendix

Most relevant PHA monomers according to Rijk et al.[317] and how they have been produced (reference retrieved with SciFinder Scholar 2006® and from Steinbüchel et al.[13])

Hydroxycarboxylic acid (CAS No.)	Production method	Reference
(R)-3-hydroxybutyrate, (R)-3-hydroxybutyric acid (141-80-0, 625-72-9)	recombinant *E. coli* HB101 with *Ralstonia eutropha* & *Pseudomonas putida* genes	Production of hydroxyalkanoate monomers by microbial fermentation. Wu, Q.; Zheng, Z.; Xi, J.; Gao, H.; Chen, G. J. Chem. Engin. Jpn. (2003), 36 (10), 1170-1173.
	Saccharomyces cerevisiae, Hansenula sp. and *Dekkera sp*	Microbiological enantioselective reduction of ethyl acetoacetate. Ribeiro, J. B.; Ramos, M. V.; de Aquino N. F. R.; Leite, S. G. F.; Antunes, O. A. C. J. Mol. Catal. B: Enzym. (2003), 24-25, 121-124.
	screening of microorganisms	Hydrolytic activity of microorganisms for kinetic resolution of a racemic ethyl 3-hydroxybutyrate. Konovalov, A. A.; Petukhova, N. I.; Ishmuratov, G. Yu.; Kharisov, R. Y.; Zorin, V. V. Bash. Khim. Z. (2000), 7(5), 34-36.
	Acetobacter & *Gluconobacter*	Acetic acid bacteria as enantioselective biocatalysts. Romano, A.; Gandolfi, R.; Nitti, P.; Rollini, M.; Molinari, F. J. Mol. Catal. B: Enzym (2002), 17(6), 235-240.
	organic synthesis	Selective reductions. Effective intramolecular asymmetric reductions of α-, β-, and γ-keto acids with diisopinocampheylborane and intermolecular asymmetric reductions of the corresponding esters with β-chlorodiisopinocampheylborane. Ramachandran, P. V.; Pitre, S; Brown, H C. J. Org. Chem. (2002), 67(15), 5315-5319.
	organic synthesis	Efficient intramolecular asymmetric reductions of α-, β-, and γ-keto acids with diisopinocampheylborane. Ramachandran, P. V; Brown, H C.; Pitre, S. Org. Lett. (2001), 3(1), 17-18.
	PHA + acid	Direct degradation of the biopolymer poly[(R)-3-hydroxybutyric acid] to (R)-3-hydroxybutanoic acid and its methyl ester. Seebach, D.; Beck, A. K.; Breitschuh, R.; Job, K. Org. Synt. (1993), 71 39-47.
	lipase	Biocatalytic resolution of (±)-hydroxyalkanoic esters. A strategy for enhancing the enantiomeric specificity of lipase-catalyzed ester hydrolysis. Scilimati, A.; Ngooi, T. K.; Sih, Charles J. Tetrahedron Lett. (1988), 29(39), 4927-30.
	organic synthesis	Synthesis of 1,3-dioxin-4-ones and their use in synthesis. Photocycloaddition, methylation, and catalytic reduction of chiral spirocyclic 1,3-dioxin-4-ones: Different stereofacial selectivity and its explanation. Sato, M.; Takayama, K.; Furuya, T.; Inukai, N.; Kaneko, C. Chem. Pharm. Bull. (1987), 35(9), 3971-4.
	organic synthesis	Synthesis of (S)- and (R)-3-hydroxy acids using cells or purified (S)-3-hydroxycarboxylate oxidoreductase from *Clostridium tyrobutyricum* and the NADP(H) regeneration system of *Clostridium thermoaceticum*. Bayer, M.; Schulz, M.; Guenther, M.; Simon, H. Appl. Microbiol. Biotechn. (1994), 42(4), 543-7.
	C. rugosa KT 8202	D-β-Hydroxyalkanoic acid. Hasegawa, J.; Ogura, M.; Kanema, H.; Kawaharada, H.; Watanabe, K. Eur. Pat. Appl. (1983), 20 pp.
	Bacilli	Products of dehydration and of polymerization of β-hydroxybutyric acid. Lemoigne, M. Bull. de la Soc. de Chim. Biol. (1926), 8, 770-82.
	C. rugosa IFO 0750	Production of D-β-hydroxycarboxylic acids from the corresponding carboxylic acids by a mutant of *Candida rugosa*. Hasegawa, J.; Ogura, M.; Kanema, H.; Kawaharada, H.; Watanabe, K. J. Ferment. Techn. (1983), 61(1), 37-42.

	sewage sludge	Combined determination of poly-β-hydroxyalkanoic and cellular fatty acids in starved marine bacteria and sewage sludge by gas chromatography with flame ionization or mass spectrometry detection. Odham, G.; Tunlid, A.; Westerdahl, G.; Maarden, P. Appl. Environ. Microbiol. (1986), 52(4), 905-10.
	anaerobic/aerobic biological phosphate removal process from sludge	Uptake of organic substrates and accumulation of polyhydroxyalkanoates linked with glycolysis of intracellular carbohydrates under anaerobic conditions in the biological excess phosphate removal process. Satoh, H.; Mino, T.; Matsuo, T. Water Sci. and Techn. (1992), 26(5-6), 933-42.
	Environmental samples and Bacillus megaterium	White polymeric β-hydroxyalkanoaltes from environmental samples and *Bacillus megaterium*. Findlay R.H., Appl. Environ. Microbial. (1983) 45 (1) 71-78.
	activated sludge	Poly-β-hydroxyalkanoate from activated sludge, Wallen L.L.; Rohwedde W.K. Environ. Sci. Techn. (1974) 8 (6) 576-579.
	Candida sp. with racemic 1,3-butanediol	Microbial manufacture of optically active 3-hydroxybutyric acid. Ito, M.; Matsuyama, A.; Kobayashi, Y. Daicel Chemical Industries, Ltd., Jpn (1991), 6 pp.
		Production of polyhydroxybutyrate by recombinant *Corynebacterium glutamicum* strain ATCC 13869. Taguchi, S.; Jo, S.-J. Agribioindustry Inc., Jpn. U.S. Pat. Appl. Publ. (2007), 20pp.
	hydrolytic enzymes from archaeobacterium *Ferroplasma acidiphilum*	Cloning, sequences and characterization of Fe^{2+}-dependent acidophilic esterase, glycosidases and DNA ligase from *Ferroplasma acidiphilum*. Golyshina, O.; Golyshina, P.; Timmis, K.; Ferrer, M. PCT Int. Appl. (2006), 35pp.
	PhaC from *Allochromatium vinosum*	In vitro analysis of the chain termination reaction in the synthesis of Poly-(*R*)-β-hydroxybutyrate by the class III synthase from *Allochromatium vinosum*. Lawrence, A. G.; Choi, J.; Rha, C.; Stubbe, J.; Sinskey, A. J. Biomacromolecules (2005), 6(4), 2113-2119.
	organic synthesis (titanium mediated aldol reaction of acyloxazolidinone)	Assembly intermediates in polyketide biosynthesis: Enantioselective syntheses of β-hydroxycarbonyl compounds. Le Sann, C.; Munoz, D. M.; Saunders, N.; Simpson, T. J.; Smith, D. I.; Soulas, F.; Watts, P.; Willis, C. L. Org. Biomol. Chem. (2005), 3(9), 1719-1728.
	P. putida NRRL B-778	The synthesis of short- and medium-chain-length poly(hydroxyalkanoate) mixtures from glucose- or alkanoic acid-grown *Pseudomonas oleovorans*. Ashby, R. D.; Solaiman, D. K. Y.; Foglia, T. A, J. Indust. Microbiol. Biotechn. (2002), 28(3), 147-153.
	PHB producers at low pH	Chiral compounds from bacterial polyesters: Sugars to plastics to fine chemicals. Lee, S. Y.;Lee, Y.;Wang, F. L. Biotechnol. Bioeng., (1999), 65 (3), 363-368.
	Bacterial excretion	Excretion of metabolites of hydrogen bacteria .3. D(-)-3-hydroxybutanoate. Vollbrecht, D.;Schlegel, H. G. Europ. J. of Appl. Microbiol. Biotech. 7(3) 259-266
(*R*)-3-hydroxypentanoic acid (53538-53-7)	*C. rugosa* KT 8202	D-β-hydroxyalkanoic acid. Hasegawa, J.; Ogura, M.; Kanema, H.; Kawaharada, H.; Watanabe, K. Eur. Pat. Appl. (1983), 20 pp.
	C. rugosa IFO 0750	Production of D-β-hydroxycarboxylic acids from the corresponding carboxylic acids by a mutant of *Candida rugosa*. Hasegawa, J.; Ogura, M.; Kanema, H.; Kawaharada, H.; Watanabe, K. J. Ferment. Technol. (1983), 61(1), 37-42.
	anaerobic/aerobic biological phosphate removal process from sludge	Uptake of organic substrates and accumulation of polyhydroxyalkanoates linked with glycolysis of intracellular carbohydrates under anaerobic conditions in the biological excess phosphate removal process. Satoh, H.; Mino, T.; Matsuo, T. Water Sci. Techn. (1992), 26(5-6), 933-42.
	Environmental samples and Bacillus megaterium	White polymeric β-hydroxyalkanoaltes from environmental samples and *Bacillus megaterium*. Findlay R.H., Appl. Environ. Microbial. (1983) 45 (1), 71-78.
	activated sludge	Poly-β-hydroxyalkanoate from activated sludge, Wallen L.L., Rohwedde W.K., Environ. Sci. Techn. (1974) 8 (6): 576-579.
		β-hydroxybutyrate polymers. Holmes P.A., Wright L.F., Collins S. H., Eur. Patent Appl. 0052459. (1981).

(R)-3-hydroxyhexanoic acid (77877-35-1)	organic synthesis (titanium mediated aldol reaction of acyloxazolidinone)	Assembly intermediates in polyketide biosynthesis: Enantioselective syntheses of β-hydroxycarbonyl compounds. Le Sann, C.; Munoz, D. M.; Saunders, N. Org. Biomol. Chem. (2005), 3(9), 1719-1728.
	P. putida	Production of chiral R-3-hydroxyalkanoic acids and R-3-hydroxyalkanoic acid methyl esters via hydrolytic degradation of polyhydroxyalkanoate synthesized by pseudomonads. De Roo, G.; Kellerhals, M. B.; Ren, Q.; Witholt, B.; Kessler, B. Biotechn. Bioeng. (2002), 77(6), 717-722.
	organic synthesis	Functionalization of unsaturated amides: Synthesis of chiral α- or β-hydroxy acids. Cardillo, G.; Hashem, M. A.; Tomasini, C. J. Chem. Soc. Perkin Transactions 1: Org. Bio-Org. Chem. (1990), (5), 1487-8.
	C. rugosa KT 8202	D-β-Hydroxyalkanoic acid. Hasegawa, J.; Ogura, M.; Kanema, H.; Kawaharada, H.; Watanabe, K. Eur. Pat. Appl. (1983), 20 pp.
	C. rugosa IFO 0750	Production of D-β-hydroxycarboxylic acids from the corresponding carboxylic acids by a mutant of Candida rugosa. Hasegawa, J.; Ogura, M.; Kanema, H.; Kawaharada, H.; Watanabe, K. J. Ferment. Technol. (1983), 61(1), 37-42.
	organic synthesis	Enantioselective aldol condensations. 2. Erythro-selective chiral aldol condensations via boron enolates. Evans, D. A.; Bartroli, J.; Shih, T. L. J. Am. Chem. Soc. (1981), 103(8), 2127-09.
	anaerobic/aerobic biological phosphate removal process from sludge	Uptake of organic substrates and accumulation of polyhydroxyalkanoates linked with glycolysis of intracellular carbohydrates under anaerobic conditions in the biological excess phosphate removal process. Satoh, H.; Mino, T.; Matsuo, T. Water Sci. and Techn. (1992), 26(5-6), 933-42.
	P. putida	Formation of polyesters by Pseudomonas oleovorans: Effect of substrates on formation and composition of poly-(R)-3-hydroxyalkanoates and poly-(R)-3-hydroxyalkenoates. Lageveen, R. G.; Huisman, G. W.; Preusting, H.; Ketelaar, P.; Eggink, G.; Witholt, B. Appl. Environ. Microbiol. (1988), 54(12), 2924-32.
	activated sludge	Poly-β-hydroxyalkanoate from activated sludge, Wallen L.L.; Rohwedde W.K. Environ. Sci. Techn. (1974), 8 (6), 576-579.
(R)-3-hydroxyheptanoic acid (85233-44-9)	nitrilase, enantioselective hydrolysis hydroxynitrile lactone biosynthesis	A mild biosynthesis of lactones via enantioselective hydrolysis of hydroxynitriles. Pollock, J. A.; Clark, K. M.; Martynowicz, B. J.; Pridgeon, M. G.; Rycenga, M. J.; Stolle, K. E.; Taylor, S. K. Tetrahedron: Asymmetry (2007), 18(16), 1888-1892.
	Cinchonidine	Hidden stereospecificity in the biosynthesis of divinyl ether fatty acids. Hamberg, M. FEBS J. (2005), 272(3), 736-743.
	C. rugosa KT 8202	D-β-Hydroxyalkanoic acid. Hasegawa, J.; Ogura, M.; Kanema, H.; Kawaharada, H.; Watanabe, K. Eur. Pat. Appl. (1983), 20 pp.
	C. rugosa IFO 0750	Production of D-β-hydroxycarboxylic acids from the corresponding carboxylic acids by a mutant of Candida rugosa. Hasegawa, J.; Ogura, M.; Kanema, H.; Kawaharada, H.; Watanabe, K. J. Ferment. Technol. (1983), 61(1), 37-42.
	P. putida	Formation of polyesters by Pseudomonas oleovorans: Effect of substrates on formation and composition of poly-(R)-3-hydroxyalkanoates and poly-(R)-3-hydroxyalkenoates. Lageveen, R. G.; Huisman, G. W.; Preusting, H.; Ketelaar, P.; Eggink, G.; Witholt, B. Appl. Environ. Microbiol. (1988), 54(12), 2924-32.
	Environmental samples and Bacillus megaterium	White polymeric β-hydroxyalkanoaltes from environmental samples and Bacillus megaterium. Findlay R.H. Appl. Environ. Microbial. (1983) 45 (1), 71-78.
(R)-3-hydroxyoctanoic acid (33825-00-2, 44987-72-6)	P. putida	Bacterial poly(hydroxyalkanoates) as a source of chiral hydroxyalkanoic acids. Ren, Q.; Grubelnik, A.; Hörler, M.; Ruth, K.; Hartmann, R.; Felber, H.; Zinn, M. Biomacromolecules (2005), 6(4), 2290-2298.
	P. putida	Production of chiral R-3-hydroxyalkanoic acids and R-3-hydroxyalkanoic acid methyl esters via hydrolytic degradation of polyhydroxyalkanoate synthesized by pseudomonads. De Roo, G.; Kellerhals, M. B.; Ren, Q.; Witholt, B.; Kessler, B. Biotechn. Bioeng. (2002), 77(6), 717-722.

	S. cerevisiae	Strategies for controlling the stereochemical course of yeast reductions. Sih, C. J.; Zhou, B. N.; Gopalan, A. S.; Shieh, W. R.; Vanmiddlesworth, F. Editor(s): Bartmann, W.; Trost, B. M. Proc. Workshop Conf. Hoechst, 14th (1984), 251-61.
	recrystallization	Preparation of optically pure 3-hydroxyalkanoic acids as intermediates for drugs and agrochemicals. Kikukawa, T.; Iizuka, Y.; Tai, A. Muraki Buhin Co., Ltd., Jpn. (1989), 5 pp.
	P. putida	Characterization of intracellular inclusions formed by Pseudomonas oleovorans during growth on octane. De Smet, M. J.; Eggink, G.; Witholt, B.; Kingma, J.; Wynberg, H. J. Bacteriol. (1983), 154(2), 870-8.
	anaerobic/aerobic biological phosphate removal process from sludge	Uptake of organic substrates and accumulation of polyhydroxyalkanoates linked with glycolysis of intracellular carbohydrates under anaerobic conditions in the biological excess phosphate removal process. Satoh, H.; Mino, T.; Matsuo, T. Water Sci. and Techn. (1992), 26(5-6), 933-42.
	Environmental samples and Bacillus megaterium	White polymeric β-hydroxyalkanoaltes from environmental samples and Bacillus megaterium. Findlay R.H. Appl. Environ. Microbial. (1983), 45 (1), 71-78.
(R)-3-hydroxynonanoic acid (33796-87-1)	organic synthesis	Enantioselective synthesis of β-hydroxy carboxylic acids: Direct conversion of β-oxocarboxylic acids to enantiomerically enriched β-hydroxy carboxylic acids via neighboring group control. Wang, Z.; Zhao, C.; Pierce, M. E.; Fortunak, J. M. Tetrahedron: Asymmetry (1999), 10(2), 225-228.
	P. putida	Formation of polyesters by Pseudomonas oleovorans: Effect of substrates on formation and composition of poly-(R)-3-hydroxyalkanoates and poly-(R)-3-hydroxyalkenoates. Lageveen, R. G.; Huisman, G. W.; Preusting, H.; Ketelaar, P.; Eggink, G.; Witholt, B. Appl. Environ. Microbiol. (1988), 54(12), 2924-32.
(R)-3-hydroxydecanoic acid (19525-80-5)	organic synthesis	Microwave-assisted cleavage of Weinreb amide for carboxylate protection in the synthesis of a (R)-3-hydroxydecanoic acid. Jaipuri, F. A.; Jofre, M. F.; Schwarz, K. A.; Pohl, N. L. Tetrahedron Letters (2004), 45(21), 4149-4152.
	E. coli HB101 with phaG from P. putida	Production of hydroxyalkanoate monomers by microbial fermentation. Wu, Q.; Zheng, Z.; Xi, J.; Gao, H.; Chen, G. J. Chem. Engin. Jpn. (2003), 36(10), 1170-1173.
	P. putida	Production of chiral R-3-hydroxyalkanoic acids and R-3-hydroxyalkanoic acid methyl esters via hydrolytic degradation of polyhydroxyalkanoate synthesized by pseudomonads. De Roo, G.; Kellerhals, M. B.; Ren, Q.; Witholt, B.; Kessler, B., Biotechn. Bioeng. (2002), 77(6), 717-722.
	organic synthesis	Enantio- and anti-diastereoselective aldol additions of acetates and propionates via O-silylketene acetals. Helmchen, G.; Leikauf, U.; Taufer-Knoepfel, I. Angew. Chem. (1985), 97(10), 874-6.
	organic synthesis	The preparation of optically pure 3-hydroxyalkanoic acid. The enantioface-differentiating hydrogenation of the carbon-oxygen double bond with modified Raney nickel. Nakahata, M.; Imaida, M.; Ozaki, H.; Harada, T.; Tai, A. Bull. Chem. Soc. Jpn. (1982), 55(7), 2186-9.
	P. putida	Formation of polyesters by Pseudomonas oleovorans: Effect of substrates on formation and composition of poly-(R)-3-hydroxyalkanoates and poly-(R)-3-hydroxyalkenoates. Lageveen, R. G.; Huisman, G. W.; Preusting, H.; Ketelaar, P.; Eggink, G.; Witholt, B. Appl. Environ. Microbiol. (1988), 54(12), 2924-32.
(R)-3-hydroxy-undecanoic acid (97961-62-1)	organic synthesis	Asymmetric synthesis via acetal templates. 12. Highly diastereoselective coupling reactions with a ketene acetal. An efficient, asymmetric synthesis of R-(+)-α-lipoic acid. Elliott, J. D.; Steele, J.; Johnson, W. S. Tetrahedron Lett. (1985), 26(21), 2535-8.
	P. putida	Sulfur containing polyhydroxyalkanoate compositions and method of production. Steinbüchel, A.; Lutke-Eversloh, T.; Ewering, C. PCT Int. Appl. (2002), 41 pp. Europ. Patent
	organic synthesis	An efficient method to chiral β-hydroxy acids: Synthesis of lipid A side chain. Nandanan, E.; Phukan, P.; Sudalai, A. Indian J. Chem., Section B: Org. Chem. incl. Med. Chem. (1999), 38B(8), 893-896.

	P. putida	Formation of polyesters by Pseudomonas oleovorans: Effect of substrates on formation and composition of poly-(R)-3-hydroxyalkanoates and poly-(R)-3-hydroxyalkenoates. Lageveen, R. G.; Huisman, G. W.; Preusting, H.; Ketelaar, P.; Eggink, G.; Witholt, B. Appl. Environ. Microbiol. (1988), 54(12), 2924-32.
(R)-3-hydroxydodecanoic acid (28254-78-6)	organic synthesis (Sharpless asym. Dihydroxylation)	Studies towards lipid A: A synthetic strategy for the enantioselective preparation of 3-hydroxy fatty acids. Guaragna, A.; De Nisco, M.; Pedatella, S.; Palumbo, G. Tetrahedron: Asymmetry (2006), 17(20), 2839-2841.
	organic synthesis (regioselective reductive opening of a cyclic sulfate)	Neurotrophic peptide aldehydes: Solid phase synthesis of fellutamide B and a simplified analog. Schneekloth, J. S.; Sanders, J. L.; Hines, J.; Crews, C. M. Bioorg. Med. Chem. Lett. (2006), 16(14), 3855-3858.
	P. putida; methanolysis	Production of chiral R-3-hydroxyalkanoic acids and R-3-hydroxyalkanoic acid methyl esters via hydrolytic degradation of poly-hydroxyalkanoate synthesized by pseudomonads. De Roo, G.; Kellerhals, M. B.; Ren, Q.; Witholt, B.; Kessler, B., Biotechn. Bioeng. (2002), 77(6), 717-722.
	P. putida	Sulfur containing polyhydroxyalkanoate compositions and method of production. Steinbüchel, A.; Lutke-Eversloh, T.; Ewering, C. PCT Int. Appl. (2002), 41 pp. Europ. Patent
	genetically engineered P. putida	Production of polyhydroxyalkanoates from intact triacylglycerols by genetically engineered pseudomonas. Solaiman, D. K. Y.; Ashby, R. D.; Foglia, T. A. Appl. Microbiol. Biotechn. (2001), 56(5-6), 64-669.
	P. mendocina 0806	Production of polyesters consisting of medium chain length 3-hydroxyalkanoic acids by Pseudomonas mendocina 0806 from various carbon sources. Tian, W.; Hong, K.; Chen, G.; Wu, Q.; Zhang, R.; Huang, W. A. v. Leeuwenhoek (2000), 77(1), 31-36.
	various microorganisms	A method for producing hydroxycarboxylic acids by auto-degradation of polyhydroxyalkanoates. Lee, S. Y.; Wang, F.; Lee, Y. LG Chemical Ltd., S. Korea. PCT Int. Appl. (1999), 68 pp.
	organic synthesis	The preparation of optically pure 3-hydroxyalkanoic acid. The enantioface-differentiating hydrogenation of the carbon-oxygen double bond with modified Raney nickel. Nakahata, M.; Imaida, M.; Ozaki, H.; Harada, T.; Tai, A. Bull. Chem. Soc. Jpn. (1982), 55(7), 2186-9.
	P. putida	Formation of polyesters by Pseudomonas oleovorans: Effect of substrates on formation and composition of poly-(R)-3-hydroxyalkanoates and poly-(R)-3-hydroxyalkenoates. Lageveen, R. G.; Huisman, G. W.; Preusting, H.; Ketelaar, P.; Eggink, G.; Witholt, B. Appl. Environ. Microbiol. (1988), 54(12), 2924-32.
(R)-3-hydroxytridecanoic acid (14664-26-7)		no relevant articles
(R)-3-hydroxytetradecanoic acid (28715-21-1)	organic synthesis	Total synthesis of (-)-tetrahydrolipstatin by the tandem Mukaiyama-aldol lactonization. Yin, J.; Yang, X. B.; Chen, Z. X.; Zhang, Y. H. Chin. Chem. Lett. (2005), 16(11), 1448-1450.
	organic synthesis	New synthesis of glycolipid immunostimulants RC-529 and CRX-524. Bazin, H. G.; Bess, L. S.; Livesay, M. T.; Ryter, K. T.; Johnson, C. L.; Arnold, J. S.; Johnson, D. A. Tetrahedron Lett. (2006), 47(13), 2087-2092.
	organic synthesis	A process in need is a process indeed: Scalable enantioselective synthesis of chiral compounds for the pharmaceutical industry. Ikunaka, M. Chem. A Eur. J. (2003), 9(2), 378-388.
	organic synthesis	Preparation of hydroxy fatty acid amides and algaecides. Kawabata, Y.; Shizusato, Y.; Tomosawa, T.; Kawamata, M.; Ueno, S. Marine Biotechnology Institute, Japan; Taisei Corp. (2002), 22 pp. patent
	genetically engineered P. putida	Production of polyhydroxyalkanoates from intact triacylglycerols by genetically engineered pseudomonas. Solaiman, D. K. Y.; Ashby, R. D.; Foglia, T. A. Appl. Microbiol. Biotechn. (2001), 56(5-6), 664-669.
	organic synthesis	An efficient synthesis of (R)-3-hydroxytetradecanoic acid. Huang, G.; Hollingsworth, R. I. Tetrahedron: Asymmetry (1998), 9(23), 4113-4115.

	S. cerevisiae	Manufacture of optically active 3-hydroxytetradecanoic acid esters. Mochizuki, N.; Oota, H.; Sukai, T. (Asahi Breweries Ltd, Japan (1994), 4 pp.
	S. cerevisiae	Chemoenzymic preparation of optically active long-chain 3-hydroxyalkanoates. Feichter, C.; Faber, K.; Griengl, H. Biocatalysis (1990), 3(1-2), 145-58.
	S. cerevisiae	Asymmetric reduction of aliphatic short- to long-chain β-keto acids by use of fermenting bakers' yeast. Utaka, M.; Watabu, H.; Higashi, H.; Sakai, T.; Tsuboi, S.; Torii, S. J. Org. Chem. (1990), 55(12), 3917-21.
	organic synthesis (reduction with NaBH$_4$)	Optically active higher 3-hydroxy fatty acids. Nakamoto, S.; Achinami, K.; Ikeda, K. Fuji Pharmaceutical Industries Co., Ltd., Japan (1986), 4 pp.
	lipase catalyst, Pseudomonas	Towards the chemoenzymic synthesis of lipid A. Sugai, T.; Ritzen, H.; Wong, C. H. Tetrahedron: Asymmetry (1993), 4(5), 1051-8.
	Pseudomonas sp. A33 and other aerobic bacteria	Biosynthesis of copolyesters consisting of 3-hydroxybutyric acid and medium-chain-length 3-hydroxyalkanoic acids from 1,3-butanediol or from 3-hydroxybutyrate by Pseudomonas sp. A33. Lee, E. Y.; Jendrossek, D.; Schirmer, A.; Choi, C. Y.; Steinbüchel, A. Appl. Microbiol. Biotechn. (1995), 42(6), 901-9.
(R)-3-hydroxypentadecanoic acid (61365-61-5)	C. antarctica lipase	Solid-phase synthesis of surfactin, a powerful biosurfactant produced by Bacillus subtilis, and of four analogues. Pagadoy, M.; Peypoux, F.; Wallach, J. Int. J. Peptide Res. Therapeut. (2005), 11(3), 195-202.
(R)-3-hydroxyhexadecanoic acid (20595-04-4)	organic synthesis	The synthesis of pseudomycin C' via a novel acid promoted side-chain deacylation of pseudomycin A. Rodriguez, M. J.; Belvo, M.; Morris, R.; Zeckner, D. J.; Current, W. L.; Sachs, R. K.; Zweifel, M. J. Bioorg. Med. Chem. Lett. (2001), 11(2), 161-164.
	organic synthesis	Preparation of hydroxy fatty acid amides and algaecides. Kawabata, Y.; Shizusato, Y.; Tomosawa, T.; Kawamata, M.; Ueno, S. Jpn. Kokai Tokkyo Koho (2002), 22 pp.
	organic synthesis	An efficient method to chiral β-hydroxy acids: Synthesis of lipid-A side chain. Nandanan, E.; Phukan, P.; Sudalai, A. Indian J. Chem., Section B: Org. Chem. incl. Med. Chem. (1999), 38B(8), 893-896.
	S. cerevisiae	Synthesis of (S)- and (R)-3-hydroxyhexadecanoic acid. Jakob, B.; Voss, G.; Gerlach, H. Tetrahedron: Asymmetry (1996), 7(11), 3255-3262.
	S. cerevisiae	Asymmetric reduction of aliphatic short- to long-chain β-keto acids by use of fermenting bakers' yeast. Utaka, M.; Watabu, H.; Higashi, H.; Sakai, T.; Tsuboi, S.; Torii, S. J. Org. Chem. (1990), 55(12), 3917-21.
	Reduction with NaBH$_4$	Optically active higher 3-hydroxy fatty acids. Nakamoto, S.; Achinami, K.; Ikeda, K. Fuji Pharmaceutical Industries Co., Ltd., Japan. Jpn. Kokai Tokkyo Koho (1986), 4 pp.
	Pseudomonas sp. A33 and other aerobic bacteria	Biosynthesis of copolyesters consisting of 3-hydroxybutyric acid and medium-chain-length 3-hydroxyalkanoic acids from 1,3-butanediol or from 3-hydroxybutyrate by Pseudomonas sp. A33. Lee, E. Y.; Jendrossek, D.; Schirmer, A.; Choi, C. Y.; Steinbüchel, A. Appl. Microbiol. Biotechn. (1995), 42(6), 901-9.
(R)-3-hydroxy-4-pentenoic acid (38996-04-2)	organic synthesis	De-novo synthesis of enantiomerically pure deoxy- and aminodeoxy-furanosides. Graef, S.; Braun, M. Liebigs Annal. Chem. (1993), (10) 1091-8.
	organic synthesis	A new approach for the chemoselective debromination of chiral bromohydrins. Toward the development of a very general approach to enantiopure α-unsubstituted β-hydroxy acids. Wang, Y.-C.; Yan, T.-H. J. Org. Chem. (2000), 65(20), 6752-6755.
	Rhodopseudomonas rubrum and P. putida	Functionalized poly-β-hydroxyalkanoates produced by bacteria. Lenz, R. W.; Kim, B. W.; Ulmer, H. W.; Fritzsche, K.; Knee, E.; Fuller, R. C. NATO ASI Series, Series E: Appl. Sci. (1990), 186, 23-35.
(S)-3-hydroxy-4-pentenoic acid (38996-05-3)	organic synthesis	Total syntheses of carbohydrates. IV. 2-Deoxy-DL-, L-, and D-erythro-pentoses and related sugars. Nakaminami, G.; Shioi, S.; Sugiyama, Y.; Isemura, S.; Shibuya, M.; Nakagawa, M. Bull. Chem. Soc. Jpn. (1972), 45(8), 2624-34.

	organic synthesis	Homochiral 4-hydroxy-5-hexenoic acids and their derivatives and homologs from carbohydrates. Song, J.; Hollingsworth, R. I. Tetrahedron: Asymmetry (2001), 12(3), 387-391.
(R)-3-hydroxy-5-hexenoic acid (119003-49-5)	organic synthesis	Process for preparation of (3R,5R)-3-hydroxy-5-(2-aminoethyl)valerolactone. Chen, W.; Chen, Z. Zhejiang University, Peop. Rep. China. Faming Zhuanli Shenqing Gongkai Shuomingshu (2005), 12 pp.
	organic synthesis	Nitrile hydratase enzymes in organic synthesis: Enantioselective synthesis of the lactone moiety of the mevinic acids. Maddrell, S. J.; Turner, N. J.; Kerridge, A.; Willetts, A. J.; Crosby, J. Tetrahedron Lett. (1996), 37(33), 6001-6004.
	S. cerevisiae	Chiral synthons for the elaboration of mevinic acid analogs. Bennett, F.; Knight, D. W. Tetrahedron Lett. (1988), 29(38), 4865-8.
	S. cerevisiae	Methyl (3R)-3-hydroxyhex-5-enoate as a precursor to chiral mevinic acid analogs. Bennett, F.; Knight, D. W.; Fenton, G. J. Chem. Soc., Perkin Transactions 1: Org. Bio-Org. Chem. (1972-1999) (1991), (1), 133-40.
	P. putida	Formation of polyesters by *Pseudomonas oleovorans*: Effect of substrates on formation and composition of poly-(R)-3-hydroxyalkanoates and poly-(R)-3-hydroxyalkenoates. Lageveen, R. G.; Huisman, G. W.; Preusting, H.; Ketelaar, P.; Eggink, G.; Witholt, B. Appl. Environ. Microbiol. (1988), 54(12), 2924-32.
	P. putida	Production of unsaturated polyesters by *Pseudomonas oleovorans*. Fritzsche, K.; Lenz, R. W.; Fuller, R. C. Int. J. Bio. Macromolecules (1990), 12(2), 85-91.
(3S,4E)-3-hydroxy-4-hexenoic acid (167077-46-5)	organic synthesis	Asymmetric Aldol Type Reactions of Acetate Imide Enolates. Yan, T.-H.; Hung, A.-W.; Lee, H.-C.; Chang, C.-S.; Liu, W.-H. J. Org. Chem. (1995), 60(11), 3301-6.
	organic synthesis	Enantioselective aldol type reactions of acetate titanium enolate with a,β-unsaturated aldehyde-TiCl₄ complex. Yan, T.-H.; Hung, A.-W.; Lee, H.-C.; Liu, W.-H.; Chang, C.-S. J. Chin. Chem. Soc. (Taipei) (1995), 42(4), 691-9.
(3R, 4E)-3-hydroxy-4-hexenoic acid (167357-28-0)	organic synthesis	A new approach for the chemoselective debromination of chiral bromohydrins. Toward the development of a very general approach to enantiopure α-unsubstituted β-hydroxy acids. Wang, Y.-C.; Yan, T.-H. J. Org. Chem. (2000), 65(20), 6752-6755.
	organic synthesis	Enantioselective aldol type reactions of acetate titanium enolate with a,β-unsaturated aldehyde-TiCl₄ complex. Yan, T.-H.; Hung, A.-W.; Lee, H.-C.; Liu, W.-H.; Chang, C.-S. J. Chin. Chem. Soc. (Taipei) (1995), 42(4), 691-9.
	P. putida	Production of unsaturated polyesters by *Pseudomonas oleovorans*. Fritzsche, K.; Lenz, R. W.; Fuller, R. C. Int. J. Bio. Macromolecules (1990), 12(2), 85-91.
	organic synthesis	Asymmetric Aldol Type Reactions of Acetate Imide Enolates. Yan, T.-H.; Hung, A.-W.; Lee, H.-C.; Chang, C.-S.; Liu, W.-H. J. Org. Chem. (1995), 60(11), 3301-6.
(R)-3-hydroxy-6-heptenoic acid (112440-75-2)	P. putida	Bacterial poly(hydroxyalkanoates) as a source of chiral hydroxyalkanoic acids. Ren, Q.; Grubelnik, A.; Hörler, M.; Ruth, K.; Hartmann, R.; Felber, H.; Zinn, M. Biomacromolecules (2005), 6(4), 2290-2298.
	organic synthesis	Total synthesis of (+)-4-oxo-5,6,9,10-tetradehydro-4,5-secofuranoeremophilane-5,1-carbolactone via novel lactone construction through allene intramolecular cycloaddition. Hayakawa, K.; Nagatsugi, F.; Kanematsu, K. J. Org. Chem. (1988), 53(4), 860-3.
	organic synthesis	Strategies for controlling the stereochemical course of yeast reductions. Sih, C. J.; Zhou, B. N.; Gopalan, A. S.; Shieh, W. R.; Vanmiddlesworth, F. Editor(s): Bartmann, Wilhelm; Trost, Barry M. Sel., Goal Synth. Effic., Proc. Workshop Conf. Hoechst, 14th (1984), 251-61.
(R)-3-hydroxy-7-octenoic acid (119003-50-8)	P. putida	Formation of polyesters by *Pseudomonas oleovorans*: Effect of substrates on formation and composition of poly-(R)-3-hydroxyalkanoates and poly-(R)-3-hydroxyalkenoates. Lageveen, R. G.; Huisman, G. W.; Preusting, H.; Ketelaar, P.; Eggink, G.; Witholt, B. Appl. Environ. Microbiol. (1988), 54(12), 2924-32.

	P. putida	Production of unsaturated polyesters by *Pseudomonas oleovorans*. Fritzsche, K.; Lenz, R. W.; Fuller, R. C. Int. J. Bio. Macromolecules (1990), 12(2), 85-91.
(R)-3-hydroxy-7-methyl-6-octenoic acid (93041-01-1)	S. cerevisiae	A useful method for preparing optically active secondary alcohols: A short enantiospecific synthesis of (R)- and (S)-sulcatol. Afonso, C. M.; Barros, M. T.; Godinho, L.; Maycock, C. D. Tetrahedron Lett. (1989), 30(20), 2707-8.
	S. cerevisiae	Convenient synthesis of (S)-citronellol of high optical purity. Hirama, M.; Noda, T.; Ito, S. J. Org. Chem. (1985), 50(1), 127-9.
(R)-3-hydroxy-8-nonenoic acid (119003-51-9)	P. putida	Bacterial poly(hydroxyalkanoates) as a source of chiral hydroxyalkanoic acids. Ren, Q.; Grubelnik, A.; Hörler, M.; Ruth, K.; Hartmann, R.; Felber, H.; Zinn, M. Biomacromolecules (2005), 6(4), 2290-2298.
	P. putida	Formation of polyesters by *Pseudomonas oleovorans*: Effect of substrates on formation and composition of poly-(R)-3-hydroxyalkanoates and poly-(R)-3-hydroxyalkenoates. Lageveen, R. G.; Huisman, G. W.; Preusting, H.; Ketelaar, P.; Eggink, G.; Witholt, B. Appl. Environ. Microbiol. (1988), 54(12), 2924-32.
(R)-3-hydroxy-9-decenoic acid (119003-52-0)	P. putida	Formation of polyesters by *Pseudomonas oleovorans*: Effect of substrates on formation and composition of poly-(R)-3-hydroxyalkanoates and poly-(R)-3-hydroxyalkenoates. Lageveen, R. G.; Huisman, G. W.; Preusting, H.; Ketelaar, P.; Eggink, G.; Witholt, B. Appl. Environ. Microbiol. (1988), 54(12), 2924-32.
(R)-cis-3-hydroxy-5-decenoic acid (403476-98-2)	Stenotrophomonas maltophilia	Selective (R)-3-hydroxylation of FA by Stenotrophomonas maltophilia. Weil, K.; Gruber, P.; Heckel, F.; Harmsen, D.; Schreier, P. Lipids (2002), 37(3), 317-323.
(R)-3-hydroxy-10-undecenoic acid (198274-26-9)	P. putida	Bacterial poly(hydroxyalkanoates) as a source of chiral hydroxyalkanoic acids. Ren, Q.; Grubelnik, A.; Hörler, M.; Ruth, K.; Hartmann, R.; Felber, H.; Zinn, M. Biomacromolecules (2005), 6(4), 2290-2298.
(R)-3-hydroxy-Z5-dodecenoic acid (88456-81-9)	Pythium ultimum	(3R,5Z)-(-)-3-Hydroxy-5-dodecenoic acid, a phytotoxic metabolite of Pythium ultimum. Ichihara, A.; Hashimoto, M; Sakamura, S. Agricult. Bio. Chem. (1985), 49(7), 2207-9.
	P. putida	Production of poly(3-hydroxyalkanoates) by *P. putida* during growth on long-chain fatty acids. Eggink, G.; Van der Wal, H.; Huyberts, G. NATO ASI Series, Series E: Appl. Sci. (1990), 186, 441-4.
(R)-3-hydroxy-Z6-dodecenoic acid (403476-95-9)	Stenotrophomonas maltophilia	Selective (R)-3-hydroxylation of FA by *Stenotrophomonas maltophilia*. Weil, K.; Gruber, P.; Heckel, F.; Harmsen, D.; Schreier, P. Lipids (2002), 37(3), 317-323.
	P. putida	Production of poly(3-hydroxyalkanoates) by *P. putida* during growth on long-chain fatty acids. Eggink, G.; Van der Wal, H.; Huyberts, G. NATO ASI Series, Series E: Appl. Sci. (1990), 186, 441-4.
(3R,6Z,9Z)-3-hydroxy-6,9-dodecadienoic acid (434955-42-7)	Stenotrophomonas maltophilia	Selective (R)-3-hydroxylation of FA by *Stenotrophomonas maltophilia*. Weil, K.; Gruber, P.; Heckel, F.; Harmsen, D.; Schreier, P. Lipids (2002), 37(3), 317-323.
(R)-3-hydroxy-Z5-tetradecenoic acid (434955-44-9)	Stenotrophomonas maltophilia	Selective (R)-3-hydroxylation of FA by *Stenotrophomonas maltophilia*. Weil, K.; Gruber, P.; Heckel, F.; Harmsen, D.; Schreier, P. Lipids (2002), 37(3), 317-323.
	P. putida	Production of poly(3-hydroxyalkanoates) by *P. putida* during growth on long-chain fatty acids. Eggink, G.; Van der Wal, H.; Huyberts, G. NATO ASI Series, Series E: Appl. Sci. (1990), 186, 441-4.
(R)-3-hydroxy-2-methyl-butanoic acid (519157-60-9)	organic synthesis	New routes to chiral Evans auxiliaries by enzymatic desymmetrization and resolution strategies. Neri, C.; Williams, J. M. J. Adv. Synth. Catalysis (2003), 345(6+7), 835-848.
	organic synthesis	Racemic auxiliaries: Applications to asymmetric synthesis. Neri, C.; Williams, J. M. J. Tetrahedron Lett. (2002), 43(23), 4257-4260.
	organic synthesis	Enantiospecific synthesis of analogs of the diketide intermediate of the erythromycin polyketide synthase (PKS). Harris, R. C.; Cutter, A. L.; Weissman, K. J.; Hanefeld, U.; Timoney, M. C.; Staunton, J. J. Chem. Res., Synopses (1998), (6), 283, 1230-1247.

	Streptomyces with hybrid type I polyketide synthases.	Polyketides and their synthesis in *Streptomyces* strains transformed with hybrid type I polyketide synthases. Leadlay, P. F.; Staunton, J.; Cortes, J. (Biotica Technology Ltd., UK). PCT Int. Appl. (1998), 177 pp.
(2R,3R)-3-hydroxy-2-methyl-butanoic acid (114892-46-5)		no relevant articles
(3R)-3-hydroxy-2,2-dimethyl-butanoic acid (35304-51-9)		no relevant articles
(3R)-3-hydroxy-2-methyl-pentanoic acid (519157-61-0)	thioesterase domains from the pimaricin biosynthetic pathways	The thioesterase domain from the pimaricin and erythromycin biosynthetic pathways can catalyze hydrolysis of simple thioester substrates. Sharma, K. K.; Boddy, C. N. Bioorg. Med. Chem. Lett. (2007), 17(11), 3034-3037.
	organic synthesis	Synthesis of (2S,3R,(1R))-4H-2,3-Dihydro-6-(1-methyl-2-oxobutyl)-2,3,5-trimethylpyran-4-one a sex pheromone of *Stegobium paniceum*. Wu, J.; Kuang, X. Sichuan Daxue Xuebao, Ziran Kexueban (2000), 37(2), 232-237.
	anaerobic/aerobic biological phosphate removal process from sludge	Uptake of organic substrates and accumulation of polyhydroxyalkanoates linked with glycolysis of intracellular carbohydrates under anaerobic conditions in the biological excess phosphate removal process. Satoh, H.; Mino, T.; Matsuo, T. Water Sci. and Techn. (1992), 26(5-6), 933-42.
(R)-3-hydroxy-4-methyl-pentanoic acid (77981-87-4)	organic synthesis	Enantioselective conjugate radical addition to β-acyloxy acrylate acceptors: An approach to acetate aldol-type products. Sibi, M. P.; Zimmerman, J.; Rheault, T. Angew. Chem., Int. Ed. (2003), 42(37), 4521-4523.
	organic synthesis	A new approach for the chemoselective debromination of chiral bromohydrins. Toward the development of a very general approach to enantiopure α-unsubstituted β-hydroxy acids. Wang, Y.-C.; Yan, T.-H. J. Org. Chem. (2000), 65(20), 6752-6755.
	organic synthesis	Highly Diastereoselective Aldol Reactions with Camphor-Based Acetate Enolate Equivalents. Palomo, C.; Oiarbide, M.; Aizpurua, J. M.; Gonzalez, A.; Garcia, J. M.; Landa, C.; Odriozola, I.; Linden, A. J. Org. Chem. (1999), 64(22), 8193-8200.
	organic synthesis (stereoselective reduction via neighboring group effect)	Enantioselective synthesis of β-hydroxy carboxylic acids: Direct conversion of β-oxocarboxylic acids to enantiomerically enriched β-hydroxy carboxylic acids via neighboring group control. Wang, Z.; Zhao, C.; Pierce, M. E.; Fortunak, J. M. Tetrahedron: Asymmetry (1999), 10(2), 225-228.
	organic synthesis	Design and evaluation of a practical camphor-based methyl ketone enolate for highly stereoselective "acetate" aldol reactions. Palomo, C.; Gonzalez, A.; Garcia, J. M.; Landa, C.; Oiarbide, M.; Rodriguez, S.; Linden, A. Angew. Chem., Int. Ed. (1998), 37(1/2), 180-182.
	organic synthesis	Asymmetric aldol reactions of chiral imidazolidinone Fischer carbene complexes. Powers, T. S.; Shi, Y.; Wilson, K. J.; Wulff, W. D.; Rheingold, A. L. J. Org. Chem. (1994), 59(23), 6882-4.
	organic synthesis	Stereoselective aldol reaction of doubly deprotonated (R)-(+)-2-hydroxy-1,2,2-triphenylethyl acetate (HYTRA): (R)-3-hydroxy-4-methylpentanoic acid (pentanoic acid, 3-hydroxy-4-methyl-, (R)-). Braun, M.; Graf, S. Org. Synth. (1995), 72, 38-47.
	organic synthesis	Optically pure, crystalline acetate aldols from N-acetylbornane-10,2-sultam. Oppolzer, W.; Starkemann, C. Tetrahedron Lett. (1992), 33(18), 2439-42.
	organic synthesis	Asymmetric and anti-selective aldolizations of acetates and propionates. Oppolzer, W.; Marco-Contelles, J. Helvet Chim. Acta (1986), 69(7), 1699-703.
	organic synthesis	Enantioselective aldol condensations. 2. Erythro-selective chiral aldol condensations via boron enolates. Evans, D. A.; Bartroli, J.; Shih, T. L. J. Am. Chem. Soc. (1981), 103(8), 2127-09.
	Environmental samples and Bacillus megaterium	White polymeric β-hydroxyalkanoaltes from environmental samples and *Bacillus megaterium*. Findlay R.H., Appl. Environ. Microbial. 1983 45 (1) 71-78.

(3R,4S)-3-hydroxy-4-methyl-hexanoic acid (214273-32-2)	organic synthesis		Stereoselective reduction of α'-branched α,β-ynones. Application to the synthesis of the octalactin A ring. Bach, J.; Garcia, J. Tetrahedron Lett. (1998), 39(37), 6761-6764.
	organic synthesis		A synthetic approach to 3-hydroxy 4-substituted carboxylic acids based on the stereoselective reduction of 1-trimethylsilyl-1-alkyn-3-ones. Alemany, C.; Bach, J.; Garcia, J.; Lopez, M.; Rodriguez, A. B. Tetrahedron (2000), 56(47), 9305-9312.
	P. putida		Production of unsaturated polyesters by *Pseudomonas oleovorans*. Fritzsche, K.; Lenz, R. W.; Fuller, R. C. Int. J. Bio. Macromolecules (1990), 12(2), 85-91.
(3R)-3-hydroxy-5-methyl-hexanoic acid (132328-50-8)	organic synthesis		Synthesis and NMR analysis in solution of oligo(3-hydroxyalkanoic acid) derivatives with the side chains of alanine, valine, and leucine (β-depsides): Coming full circle from PHB to β-peptides to PHB. Albert, M.; Seebach, D.; Duchardt, E.; Schwalbe, H. Helvet Chim. Acta (2002), 85(2), 633-658.
	organic synthesis		Enantioselective synthesis of β-hydroxy carboxylic acids: Direct conversion of β-oxocarboxylic acids to enantiomerically enriched β-hydroxy carboxylic acids via neighboring group control. Wang, Z.; Zhao, C.; Pierce, M. E.; Fortunak, J. M. Tetrahedron: Asymmetry (1999), 10(2), 225-228.
	P. putida		Production of unsaturated polyesters by *Pseudomonas oleovorans*. Fritzsche, K.; Lenz, R. W.; Fuller, R. C. Int. J. Bio. Macromolecules (1990), 12(2), 85-91.
(3R)-3-hydroxy-9-methyl-decanoic acid (56221-73-9)	*P. putida*		Biosynthesis of methyl-branched PHA by *Pseudomonas oleovorans*, Hazer, B.; Lenz, R.W.; Fuller, R.C. Macromolecules (1994), 27 (1), 45-49.
(R)-4-hydroxy-pentanoic acid (155847-13-5)	*Proteus mirabilis* and *P. vulgaris*		On a non-pyridine nucleotide-dependent 2-oxo-acid reductase of broad substrate specificity from two Proteus species. Neumann, S.; Simon, H. FEBS Lett. (1984), 167(1), 29-32.
	organic synthesis		Selective reductions. Effective intramolecular asymmetric reductions of α-, β-, and γ-keto acids with diisopinocampheylborane and intermolecular asymmetric reductions of the corresponding esters with B-chlorodiisopinocampheylborane. Ramachandran, P. V.; Pitre, S.; Brown, H. C. J. Org. Chem. (2002), 67(15), 5315-5319.
	organic synthesis		Efficient intramolecular asymmetric reductions of α-, β-, and γ-keto acids with diisopinocampheylborane. Ramachandran, P. V.; Brown, H. C.; Pitre, S. Org. Lett. (2001), 3(1), 17-18.
	Alcaligenes, Paracoccus, Pseudomonas and *Methylobacterium*		Identification of 4-hydroxyvaleric acid as a constituent of biosynthetic polyhydroxyalkanoic acids from bacteria. Valentin, H. E.; Schönebaum, A.; Steinbüchel, A. Appl. Microbiol. Biotechn. (1992), 36(4), 507-14.
(R)-4-hydroxy-hexanoic acid (87241-93-8)	organic synthesis		Selective reductions. Effective intramolecular asymmetric reductions of α-, β-, and g-keto acids with diisopinocampheylborane and intermolecular asymmetric reductions of the corresponding esters with B-chlorodiisopinocampheylborane. Ramachandran, P. V.; Pitre, S.; Brown, H. C. J. Org. Chem. (2002), 67(15), 5315-5319.
	organic synthesis		Efficient Intramolecular Asymmetric Reductions of α-, β-, and g-Keto Acids with Diisopinocampheylborane. Ramachandran, P. V.; Brown, H. C.; Pitre, S. Org. Lett. (2001), 3(1), 17-18.
	recombinant strains of the PHA-neg. mutants *P. putida* GPp104 and *Alcaligenes eutrophus* PHB-4		Identification of 4-hydroxyhexanoic acid as a new constituent of biosynthetic polyhydroxyalkanoic acids from bacteria. Valentin, H. E.; Lee, E.; Choi, C.; Steinbüchel, A. Appl. Microbiol. Biotechn. (1994), 40(5), 710-16.
(R)-4-hydroxy-decanoic acid (30100-79-9)	*C. petrophilum* ATCC 20226, *C. oleophila* ATCC20177, *C. sake* ATCC 28137		Process for preparing compositions containing unsaturated lactones and organoleptic uses of the lactones. Farbood, M. I.; Morris, J. A.; Sprecker, M. A.; Bienkowski, L. J.; Miller, K. P.; Vock, M. H.; Hagedorn, M. L. International Flavors and Fragrances Inc., USA. Eur. Pat. (1990), 127 pp.
			Formation of novel PHAs from long chain fatty acids. Eggink, G.; Huijbert, G.N.M.; van der Wal, H.; de Waard, P. Lecture at the Int. Symp. on bacterial PHA, Montreal, Canada (1994).
(R)-5-hydroxy-hexanoic acid (83972-61-6)	organic synthesis		TFA-catalyzed trimerization of R-(+)-6-methyltetrahydro-2-pyranone. Fazio, F.; Schneider, M. P. Tetrahedron Lett. (2002), 43(5), 811-814.

Compound	Organism	Reference
(R)-6-hydroxy-Z3-dodecenoic acid (129830-47-3)	C. petrophilum ATCC 20226, C. oleophila ATCC20177, C. sake ATCC 28137	H.E. Valentin, A. Schönebaum and A. Steinbüchel, unpublished results Process for preparing compositions containing unsaturated lactones and organoleptic uses of the lactones. Farbood, M. I.; Morris, J. A.; Sprecker, M. A.; Bienkowski, L. J.; Miller, K. P.; Vock, M. H.; Hagedorn, M. L. International Flavors and Fragrances Inc., USA. Eur. Pat. (1990), 127 pp.
	P. aeruginosa 44T1	Formation of novel poly(hydroxyalkanoates) from long-chain fatty acids. Eggink, G.; de Waard, P.; Huijberts, G. N. M. Can. J. Microbiol. (1995), 41(Suppl. 1), 14-21.
(3R)-6-bromo-3-hydroxy-hexanoic acid (581809-21-4)	Pseudomonas bacteria (cichorii, jessenii, putida)	Polyhydroxyalkanoate copolymer including unit having bromo group in side chain and production method thereof. Honma, T.; Kozaki, S.; Imamura, T.; Kenmoku, T.; Fukui, T.; Sugawa, E.; Yano, T. Eur. Pat. Appl. (2003), 46 pp.
	R. rubrum and P. putida	Functionalized poly-β-hydroxyalkanoates produced by bacteria. Lenz, R. W.; Kim, B. W.; Ulmer, H. W.; Fritzsche, K.; Knee, E.; Fuller, R. C. NATO ASI Series, Series E: Appl. Sci. (1990), 186, 23-35.
(3R)-8-bromo-3-hydroxy-octanoic acid (581809-22-5)	Pseudomonas bacteria (cichorii, jessenii, putida)	Polyhydroxyalkanoate copolymer including unit having bromo group in side chain and production method thereof. Honma, T.; Kozaki, S.; Imamura, T.; Kenmoku, T.; Fukui, T.; Sugawa, E.; Yano, T. Eur. Pat. Appl. (2003), 46 pp.
	R. rubrum and P. putida	Functionalized poly-β-hydroxyalkanoates produced by bacteria. Lenz, R. W.; Kim, B. W.; Ulmer, H. W.; Fritzsche, K.; Knee, E.; Fuller, R. C. NATO ASI Series, Series E: Appl. Sci. (1990), 186, 23-35.
(3R)-11-bromo-3-hydroxy-undecanoic acid (222043-79-0)	R. rubrum and P. putida	Functionalized poly-β-hydroxyalkanoates produced by bacteria. Lenz, R. W.; Kim, B. W.; Ulmer, H. W.; Fritzsche, K.; Knee, E.; Fuller, R. C. NATO ASI Series, Series E: Appl. Sci. (1990), 186, 23-35.
(R)-6,6,6-trifluoro-3-hydroxy-hexanoic acid (176666-79-8)	P. oleovorans (ATCC 29347) P. putida (KT 2442)	Microbial synthesis of poly(β-hydroxyalkanoates) containing fluorinated side-chain substituents. Kim, O.; Gross, R. A.; Hammar, W. J.; Newmark, R. A. Macromolecules (1996), 29(13), 4572-4581.
(3R)-3-hydroxy-4-phenoxy-butanoic acid (393085-31-9)	organic synthesis	Process for preparing chiral 4-substituted-3-hydroxybutyric acid and its salt from (S)-3-activated hydroxybutyrolactone. Byun, I. S.; Kim, K. I.; Ha, S. B. Samsung Fine Chemical Co., Ltd., Korea. Taeho Kongbo (2000) Y. Kim, personal conversation with A. Steinbüchel (1995)
(R)-3-hydroxy-5-phenylvaleric acid (153744-07-1)	organic synthesis	Stereoselective synthesis of 2-azetidinones as cholesterol-absorption inhibitors. Annunziata, R.; Benaglia, M.; Cinquini, M.; Cozzi, F. Tetrahedron: Asymmetry (1999), 10(24), 4841-4849.
	Pseudomonas bacteria (cichorii, jessenii, putida)	Production of novel polyhydroxyalkanoate copolymers by Pseudomonas. Kenmoku, T.; Yano, T.; Mihara, C.; Kozaki, S.; Honma, T.; Fukui, T.; Imamura, T. PCT Int. Appl. (2004), 245 pp.
	P. aeruginosa, P. putida, R. eutropha	Chiral compounds from bacterial polyesters: Sugars to plastics to fine chemicals. Lee, S. Y.; Lee, Y.; Wang, F. Biotechn. Bioengin. (1999), 65(3), 363-368.
	A. latus, Bacillus (bacterium genus), Corynebacterium, Escherichia coli, Methylosinus trichosporium, P. aeruginosa, P. putida, R. eutropha, Rhodospirillum rubrum	A method for producing hydroxycarboxylic acids by auto-degradation of polyhydroxyalkanoates. Lee, S. Y.; Wang, F.; Lee, Y. LG Chemical Ltd., S. Korea. PCT Int. Appl. (1999), 68 pp.
	organic synthesis	Resolution and asymmetric synthesis of 3-hydroxy carboxylic acids by using (-)-menthone as a chiral template. Harada, T.; Yoshida, T.; Kagamihara, Y.; Oku, A. J. Chem. Soc., Chem. Comm. (1993), (17), 1367-70.
	P. putida	An unusual bacterial polyester with a phenyl pendant group. Fritzsche, K.; Lenz, R. W.; Fuller, R. C. Makromol. Chem. (1990), 191(8), 1957-65.
	P. putida	Preparation and characterization of poly(β-hydroxyalkanoates) obtained from Pseudomonas oleovorans grown with mixtures of 5-phenylvaleric acid and n-alkanoic acids. Kim, Y. B.; Lenz, R. W.; Fuller, R. C. Macromolecules (1991), 24(19), 5256-60.

Compound	Organism	Reference
(3R)-3-hydroxy-5-phenoxy-pentanoic acid (173395-00-1)	*Pseudomonas* bacteria (*cichorii, jessenii, putida*)	Production of novel polyhydroxyalkanoate copolymers by *Pseudomonas*. Kenmoku, T.; Yano, T.; Mihara, C.; Kozaki, S.; Honma, T.; Fukui, T.; Imamura, T. PCT Int. Appl. (2004), 245 pp.
	Pseudomonas bacteria (*cichorii, jessenii, putida*)	Polyhydroxyalkanoate copolymer including unit having bromo group in side chain and production method thereof. Honma, T.; Kozaki, S.; Imamura, T.; Kenmoku, T.; Fukui, T.; Sugawa, E.; Yano, T. Eur. Pat. Appl. (2003), 46 pp.
	P. putida	Biosynthesis of novel aromatic copolyesters from insoluble 11-phenoxyundecanoic acid by *Pseudomonas putida* BM01. Song, J. J.; Yoong, S. C. Appl. Environ. Microbiol. (1996), 62(2), 536-44.
		Y. Kim, personal conversation A. Steinbüchel (1995)
(3R)-3-hydroxy-benzenehexanoic acid (247169-45-5)	genetically engineered strains of *P. putida*	Genetically engineered pseudomonas: A factory of new bioplastics with broad applications. Olivera, E. R.; Carnicero, D.; Jodra, R.; Minambres, B.; Garcia, B.; Abraham, G. A.; Gallardo, A.; San Roman, J.; Garcia, J. L.; Naharro, G.; Luengo, J. M. Environ. Microbiol. (2001), 3(10), 612-618.
(3R)-3-hydroxy-benzeneheptanoic acid (247169-47-7)		no relevant articles
(R)-3-hydroxy-7-phenoxy-heptanoic acid (173395-01-2)	*P. putida*	Biosynthesis of novel aromatic copolyesters from insoluble 11-phenoxyundecanoic acid by *Pseudomonas putida* BM01. Song, J. J.; Yoong, S. C. Appl. Environ. Microbiol. (1996), 62(2), 536-44.
		Y. Kim, personal conversation A. Steinbüchel (1995)
(3R)-8-(acetyloxy)-3-hydroxy-octanoic acid (318967-06-5)	*P. putida*	Bacterial synthesis of poly-β-hydroxyalkanoates with functionalized side chains. Lenz, R. W.; Fuller, R. C.; Scholz, C.; Touraud, F. Stud. Polym. Sci. (1994), 12, 109-19.
4-hydroxy-heptanoic acid (44911-93-5)		H. E. Valentin, A. Schönebaum, A. Steinbüchel, unpublished results
4-hydroxy-octanoic acid (7779-55-7)		H. E. Valentin, A. Schönebaum, A. Steinbüchel, unpublished results
6-hydroxydodecanoic acid (35875-13-9)	*P. putida*	Formation of polyesters by *Pseudomonas oleovorans*: Effect of substrates on formation and composition of poly-(R)-3-hydroxyalkanoates and poly-(R)-3-hydroxyalkenoates. Lageveen, R. G.; Huisman, G. W.; Preusting, H.; Ketelaar, P.; Eggink, G.; Witholt, B. Appl. Environ. Microbiol. (1988), 54(12), 2924-32.
3-hydroxy-6-octenoic acid (128940-64-7)	*P. putida*	Production of unsaturated polyesters by *Pseudomonas oleovorans*. Fritzsche, K.; Lenz, R. W.; Fuller, R. C. Int. J. Bio. Macromolecules (1990), 12(2), 85-91.
(3R,7Z)-3-hydroxy-7-tetradecenoic acid (154357-32-1)	*Chromobacterium* sp.	Biosynthesis of polyhydroxyalkanoates from 1,3-propanediol by *Chromobacterium sp*. Kimura, H.; Yamamoto, T.; Iwakura, K. Polym. J. Jpn. (2002), 34(9), 659-665.
	P. putida KT2442	^{13}C nuclear magnetic resonance studies of *Pseudomonas putida* fatty acid metabolic routes involved in poly(3-hydroxyalkanoate) synthesis. Huijberts, G. N. M.; de Rijk, T.; de Waard, P.; Eggink, G. J. Bacteriol. (1994), 176(6), 1661-6.
	P. citronellolis	Polyester biosynthesis characteristics of *Pseudomonas citronellolis* grown on various carbon sources, including 3-methyl-branched substrates. Choi, M. H.; Yoon, S. C. Appl. Environ. Microbiol. (1994), 60(9), 3245-54.
(3R,5Z,8Z)-3-hydroxy-5,8-tetradecadienoic acid (202601-25-0)	organic synthesis (Lindlar's catalyst/enzymatic(soybean LOX-1))	A novel synthesis of 3(R)-HETE, 3(R)-HTDE and enzymatic synthesis of 3(R),15(S)-DiHETE. Groza, N. V.; Ivanov, I. V.; Romanov, S. G.; Myagkova, G. I.; Nigam, S. Tetrahedron (2002), 58(49), 9859-9863.
	Stenotrophomonas maltophilia	Selective (R)-3-hydroxylation of FA by *Stenotrophomonas maltophilia*. Weil, K.; Gruber, P.; Heckel, F.; Harmsen, D.; Schreier, P. Lipids (2002), 37(3), 317-323.

Compound	Source	Reference
	yeast *Dipodascopsis uninucleata* UOFS Y 128	Production of 3*R*-hydroxy-polyenoic fatty acids by the yeast *Dipodascopsis uninucleata*. Venter, P.; Kock, J. L. F.; Kumar, G. Sravan; Botha, A.; Coetzee, D. J.; Botes, P. J.; Bhatt, R. K.; Falck, J. R.; Schewe, T.; Nigam, S. Lipids (1997), 32(12), 1277-1283.
	P. putida	Production of poly(3-hydroxyalkanoates) by *P. putida* during growth on long-chain fatty acids. Eggink, G.; Van der Wal, H.; Huyberts, G. NATO ASI Series, Series E: Appl. Sc. (1990), 186, 441-4.
3-hydroxy-6-methyl-heptanoic acid (230949-26-5)	Environmental samples and *Bacillus megaterium*	White polymeric β-hydroxyalkanoaltes from environmental samples and *Bacillus megaterium*. Findlay R.H. Appl. Environ. Microbial. (1983), 45 (1), 71-78.
(3R,5R)-3-hydroxy-5-methyl-octanoic acid (43209-54-7)		Preparation of β-amino acids having affinity for the α-2-d protein. Conway, B. G.; Nanninga, T. N.; Wu, H.; Hoge, G.; Pearlman, B. A.; Winkle, D. D. Warner-Lambert Company LLC, PCT Int. Appl. (2006), 88 pp.
	P. putida	Bacterial polyesters containing branched poly(β-hydroxyalkanoate) units. Fritzsche, K.; Lenz, R. W.; Fuller, R. C. Int. J. Bio. Macromolecules (1990), 12(2), 92-101.
(*R*)-3-hydroxy-6-methyl-octanoic acid (59896-39-8)	*P. putida*	Bacterial polyesters containing branched poly(β-hydroxyalkanoate) units. Fritzsche, K.; Lenz, R. W.; Fuller, R. C. Int. J. Bio. Macromolecules (1990), 12(2), 92-101.
(*R*)-3-hydroxy-7-methyl-octanoic acid (634602-29-2)	Environmental samples and Bacillus megaterium	White polymeric β-hydroxyalkanoaltes from environmental samples and *Bacillus megaterium*. Findlay R.H. Appl. Environ. Microbial. (1983), 45 (1), 71-78.
	P. putida	Bacterial polyesters containing branched poly(β-hydroxyalkanoate) units. Fritzsche, K.; Lenz, R. W.; Fuller, R. C. Int. J. Bio. Macromolecules (1990), 12(2), 92-101.
(*R*)-3-hydroxy-6-methyl-nonanoic acid (257301-18-1)	*P. putida*	Biosynthesis of methyl-branched PHA by *Pseudomonas oleovorans*, Hazer, B.; Lenz, R.W.; Fuller, R.C. Macromolecules (1994), 27 (1), 45-49.
3-hydroxy-7-methyl-nonanoic acid (122751-73-9)	*P. putida*	Biosynthesis of methyl-branched PHA by *Pseudomonas oleovorans*, Hazer, B.; Lenz, R.W.; Fuller, R.C. Macromolecules (1994), 27 (1), 45-49.
(*R*)-3-hydroxy-8-methyl-nonanoic acid (62675-78-9)	*P. putida*	Biosynthesis of methyl-branched PHA by *Pseudomonas oleovorans*, Hazer, B.; Lenz, R.W.; Fuller, R.C. Macromolecules (1994), 27 (1), 45-49.
(2S)-hydroxy-butanedioic acid-1-methyl ester (66212-45-1)	organic synthesis	Efficient synthesis of N-benzyloxycarbonyl- and N-tert-butoxycarbonyl-(S)-isoserine and their derivatives. Andruszkiewicz, R.; Wyszogrodzka, M. Synlett (2002), (12), 2101-2103.
	organic synthesis	The diastereoselective asymmetric total synthesis of NG-391, a neuronal cell-protecting molecule. Hayashi, Y.; Yamaguchi, J.; Shoji, M. Tetrahedron (2002), 58(49), 9839-9846.
	C. rugosa	Manufacture of (2S,3R)-methylmalic acid. Asada, M.; Sawa, I.; Hasegawa, J.; Watanabe, K. (Kanegafuchi Chemical Industry Co., Ltd.,). Jpn. Kokai Tokkyo Koho (1986), 6 pp.
(3R)-3-hydroxy-hexanedioic acid-6-methyl ester (647831-66-1)		C. Scholz, personal communication to A. Steinbüchel
(3R)-3-hydroxy-octanedioic acid-8-methyl ester (647831-62-7)	*Pseudomonas* bacteria	Production of novel polyhydroxyalkanoate copolymers by *Pseudomonas*. Kenmoku, T.; Yano, T.; Mihara, C.; Kozaki, S.; Honma, T.; Fukui, T.; Imamura, T. PCT Int. Appl. (2004), 245 pp.
	P. putida	Bacterial synthesis of poly-β-hydroxyalkanoates with functionalized side chains. Lenz, R. W.; Fuller, R. C.; Scholz, C.; Touraud, F. Stud. Polym. Sci. (1994), 12, 109-19.
(3R)-3-hydroxy-decanedioic acid-10-methyl ester (647831-63-8)	*Pseudomonas* bacteria	Production of novel polyhydroxyalkanoate copolymers by *Pseudomonas*. Kenmoku, T.; Yano, T.; Mihara, C.; Kozaki, S.; Honma, T.; Fukui, T.; Imamura, T. PCT Int. Appl. (2004), 245 pp.
	P. putida	Bacterial synthesis of poly-β-hydroxyalkanoates with functionalized side chains. Lenz, R. W.; Fuller, R. C.; Scholz, C.; Touraud, F. Stud. Polym. Sci. (1994), 12, 109-19.

(3R)-3-hydroxy-octanedioic acid-8-ethyl ester (686753-19-5)	*Pseudomonas* bacteria	Production of novel polyhydroxyalkanoate copolymers by *Pseudomonas*. Kenmoku, T.; Yano, T.; Mihara, C.; Kozaki, S.; Honma, T.; Fukui, T.; Imamura, T. PCT Int. Appl. (2004), 245 pp.
	P. putida	Bacterial synthesis of poly-β-hydroxyalkanoates with functionalized side chains. Lenz, R. W.; Fuller, R. C.; Scholz, C.; Touraud, F. Stud. Polym. Sci. (1994), 12, 109-19.
(3R)-3-hydroxy-decanedioic acid-10-ethyl ester (686753-18-4)	*Pseudomonas* bacteria	Production of novel polyhydroxyalkanoate copolymers by *Pseudomonas*. Kenmoku, T.; Yano, T.; Mihara, C.; Kozaki, S.; Honma, T.; Fukui, T.; Imamura, T. PCT Int. Appl. (2004), 245 pp.
	P. putida	Bacterial synthesis of poly-β-hydroxyalkanoates with functionalized side chains. Lenz, R. W.; Fuller, R. C.; Scholz, C.; Touraud, F. Stud. Polym. Sci. (1994), 12, 109-19.
(3R)-3-hydroxy-8-phenoxy-octanoic acid (174792-08-6)		Y. Kim, personal conversation with A. Steinbüchel (1995)
(3R)-4-(4-cyanophenoxy)-3-hydroxy-butanoic acid (484040-52-0)		Biological and chemical routes to modulate PHA structure. Gross, R. A., (1994) Lecture at Int. Symp. on bacterial PHA, Montreal, Canada.
	Pseudomonas cichorii & *jessenii*	Poly(hydroxyalkanoates) having 3-hydroxy-5-(4-cyanophenoxy)valeric acid units and their manufacture. Imamura, T.; Kenmoku, T.; Honma, T.; Sugawa, E.; Yano, T. Canon Inc. Jpn. Kokai Tokkyo Koho (2002), 13 pp.
(3R)-5-(4-cyanophenoxy)-3-hydroxy-pentanoic acid (457655-22-0)		Biological and chemical routes to modulate PHA structure. Gross, R. A., (1994) Lecture at Int. Symp. on bacterial PHA, Montreal, Canada.
(3R)-6-(4-cyanophenoxy)-3-hydroxy-hexanoic acid (164855-47-4)	*Pseudomonas* bacteria (*cichorii, jessenii, putida*)	Polyhydroxyalkanoate copolymer including unit having bromo group in side chain and production method thereof. Honma, T.; Kozaki, S.; Imamura, T.; Kenmoku, T.; Fukui, T.; Sugawa, E.; Yano, T. Eur. Pat. Appl. (2003), 46 pp.
(3R)-3-hydroxy-cyclohexanebutanoic acid (483343-33-5)	*Pseudomonas* bacteria	Production of novel polyhydroxyalkanoate copolymers by *Pseudomonas*. Kenmoku, T.; Yano, T.; Mihara, C.; Kozaki, S.; Honma, T.; Fukui, T.; Imamura, T. PCT Int. Appl. (2004), 245 pp.
	P. putida	Production of unusual bacterial polyesters by *Pseudomonas oleovorans* through cometabolism. Lenz, R. W.; Kim, Y. B.; Fuller, R. C. FEMS Microbiol. Rev. (1992), 103(2-4), 207-14.
3,12-dihydroxydodecanoic acid (80828-81-5)	*P. putida*	Production of unusual bacterial polyesters by *Pseudomonas oleovorans* through cometabolism. Lenz, R. W.; Kim, Y. B.; Fuller, R. C. FEMS Microbiol. Rev. (1992), 103(2-4), 207-14.
	P. aeruginosa 44T1	Formation of novel poly(hydroxyalkanoates) from long-chain fatty acids. Eggink, G.; de Waard, P.; Huijberts, G. N. M. Can. J. Microbiol. (1995), 41(Suppl. 1), 14-21.
[R-[R*,R*-(Z)]]-3,8-dihydroxy-5-tetradecenoic acid (165393-38-4)		Formation of novel PHAs from long chain fatty acids. Eggink, G.; Huijbert, G.N.M.; van der Wal, H.; de Waard, P. Lecture at the Int. Symp. on bacterial PHA, Montreal, Canada (1994).
	P. aeruginosa	Formation of novel poly(hydroxyalkanoates) from long-chain fatty acids. Eggink, G.; de Waard, P.; Huijberts, G. N. M. Can. J. Microbiol. (1995), 41(Suppl. 1), 14-21.
[2S-[2a(R*),3a]]-β-hydroxy-3-pentyl-oxiranepropanoic acid (165393-40-8)		Formation of novel PHAs from long chain fatty acids. Eggink, G.; Huijbert, G.N.M.; van der Wal, H.; de Waard P. Lecture at the Int. Symp. on bacterial PHA, Montreal, Canada (1994).
	P. aeruginosa	Formation of novel poly(hydroxyalkanoates) from long-chain fatty acids. Eggink, G.; de Waard, P.; Huijberts, G. N. M. Can. J. Microbiol. (1995), 41(Suppl. 1), 14-21.
[2S-[2a(S*),3a]]-β-hydroxy-3-pentyl-oxiranepentanoic acid (165393-39-5)		Formation of novel PHAs from long chain fatty acids. Eggink, G.; Huijbert, G.N.M.; van der Wal, H.; de Waard P. Lecture at the Int. Symp. on bacterial PHA, Montreal, Canada (1994).
	P. aeruginosa	Formation of novel poly(hydroxyalkanoates) from long-chain fatty acids. Eggink, G.; de Waard, P.; Huijberts, G. N. M. Can. J. Microbiol. (1995), 41(Suppl. 1), 14-21.

Compound	Organism	Reference
[2S-[2a(3S*,5Z),3a]]-3-hydroxy-7-(3-pentyloxiranyl)-5-heptenoic acid (165393-37-3)		Formation of novel PHAs from long chain fatty acids. Eggink, G.; Huijbert, G.N.M.; van der Wal, H.; de Waard, P. Lecture at the Int. Symp. on bacterial PHA, Montreal, Canada (1994).
7-fluoro-3-hydroxy-heptanoic acid (131482-58-1)	P. putida	New bacterial copolyester of 3-hydroxyalkanoates and 3-hydroxy-ω-fluoroalkanoates produced by Pseudomonas oleovorans. Abe, C.; Taima, Y.; Nakamura, Y.; Doi, Y. Polym. Comm. (1990), 31(11), 404-6.
9-fluoro-3-hydroxy-nonanoic acid (131482-59-2)	P. putida	New bacterial copolyester of 3-hydroxyalkanoates and 3-hydroxy-ω-fluoroalkanoates produced by Pseudomonas oleovorans. Abe, C.; Taima, Y.; Nakamura, Y.; Doi, Y. Polym. Comm. (1990), 31(11), 404-6.
6-chloro-3-hydroxy-hexanoic acid (128114-89-6)	R. rubrum and P. oleovorans	Functionalized poly-β-hydroxyalkanoates produced by bacteria. Lenz, R. W.; Kim, B. W.; Ulmer, H. W.; Fritzsche, K.; Knee, E.; Fuller, R. C. NATO ASI Series, Series E: Appl. Sci. (1990), 186, 23-35.
8-chloro-3-hydroxy-octanoic acid (128114-88-5)	P. putida	Biosynthesis and characterization of a new bacterial copolyester of 3-hydroxyalkanoates and 3-hydroxy-ω-chloroalkanoates. Doi, Y.; Abe, C. Macromolecules (1990), 23(15), 3705-7.
	P. aeruginosa 44T1	Formation of novel poly(hydroxyalkanoates) from long-chain fatty acids. Eggink, G.; de Waard, P.; Huijberts, G. N. M. Can. J. Microbiol. (1995), 41(Suppl. 1), 14-21.
[R-(Z)]-6-hydroxy-3-dodecenoic acid (129830-47-3)	Candida oleophila, Candida petrophilum, Candida sake	Process for preparing compositions containing unsaturated lactones and organoleptic uses of the lactones. Farbood, M. I.; Morris, J. A.; Sprecker, M. A.; Bienkowski, L. J.; Miller, K. P.; Vock, M. H.; Hagedorn, M. L. International Flavors and Fragrances Inc., USA. Eur. Pat. (1990), 127 pp. Formation of novel PHAs from long chain fatty acids. Eggink G., Huijbert G.N.M., van der Wal H., de Waard P., Lecture at the Int. Symp. on bacterial PHA, Montreal, Canada (1994).
	organic synthesis	Total synthesis of calonyctin A2, a macrolidic glycolipid with plant growth-promoting activity. Furukawa, J.-I.; Kobayashi, S.; Nomizu, M.; Nishi, N.; Sakairi, N. Tetrahedron Lett. (2000), 41(18), 3453-3457.
(2R,3R)-3-hydroxy-2-methyl-butanoic acid (114892-46-5)	anaerobic/aerobic biological phosphate removal process from sludge	Uptake of organic substrates and accumulation of polyhydroxyalkanoates linked with glycolysis of intracellular carbohydrates under anaerobic conditions in the biological excess phosphate removal process. Satoh, H.; Mino, T.; Matsuo, T. Water Sci. and Techn. (1992), 26(5-6), 933-42.
3-hydroxy-4-cis-hexenoic acid	P. putida	Production of unsaturated polyesters by Pseudomonas oleovorans. Fritzsche, K.; Lenz, R. W.; Fuller, R. C. Int. J. Bio. Macromolecules (1990), 12(2), 85-91.
3-hydroxy-4-methyl-octanoic acid	P. putida	Bacterial polyesters containing branched poly(β-hydroxyalkanoate) units. Fritzsche, K.; Lenz, R. W.; Fuller, R. C. Int. J. Bio. Macromolecules (1990), 12(2), 92-101.
3-hydroxy-7-methyl-decanoic acid	P. putida	Biosynthesis of methyl-branched PHA by Pseudomonas oleovorans, Hazer, B.; Lenz, R.W.; Fuller R.C. Macromolecules (1994), 27 (1), 45-49.
3-hydroxyazelaic acid methyl ester	P. putida	Bacterial synthesis of poly-β-hydroxyalkanoates with functionalized side chains. Lenz, R. W.; Fuller, R. C.; Scholz, C.; Touraud, F. Stud. Polym. Sci. (1994), 12, 109-19.
3-hydroypimelic acid-propyl ester		C. Scholz, personal communication to A. Steinbüchel
3-hydroxysebacic acid-benyl ester		C. Scholz, personal communication to A. Steinbüchel
3-hydroxy-8-acetoxyoctanoic acid	P. putida	Bacterial synthesis of poly-β-hydroxyalkanoates with functionalized side chains. Lenz, R. W.; Fuller, R. C.; Scholz, C.; Touraud, F. Stud. Polym. Sci. (1994), 12, 109-19.
3-hydroxy-9-acetoxynonanoic acid	P. putida	Bacterial synthesis of poly-β-hydroxyalkanoates with functionalized side chains. Lenz, R. W.; Fuller, R. C.; Scholz, C.; Touraud, F. Stud. Polym. Sci. (1994), 12, 109-19.
para-nitro-phenoxy-3-hydroxy-hexanoic acid		Biological and chemical routes to modulate PHA structure. Gross, R. A.; (1994) Lecture at Int. Symp. on bacterial PHA, Montreal, Canada.
7-cyano-3-hydroxyheptanoic acid	P. putida	Production of unusual bacterial polyesters by Pseudomonas oleovorans through cometabolism. Lenz, R. W.; Kim, Y. B.; Fuller, R. C. FEMS Microbiol. Rev. (1992), 103(2-4), 207-14.

9-cyano-3-hydroxynonanoic acid	*P. putida*	Production of unusual bacterial polyesters by *Pseudomonas oleovorans* through cometabolism. Lenz, R. W.; Kim, Y. B.; Fuller, R. C. FEMS Microbiol. Rev. (1992), 103(2-4), 207-14.
3-hydroxy-2,6-dimethyl-5-heptenoic acid	*P. putida*	Biosynthesis of methyl-branched PHA by *Pseudomonas oleovorans*. Hazer B., Lenz R.W., Fuller R.C., Macromolecules (1994) 27 (1): 45-49

Publications:

Ren, Q.; Grubelnik, A.; Hörler, M.; **Ruth, K.**; Hartmann, R.; Felber, H.; Zinn, M. Bacterial poly(hydroxyalkanoates) as a source of chiral hydroxyalkanoic acids. *Biomacromolecules* 2005, 6 (4), 2290-2298.

Furrer, P.; Hany, R.; Rentsch, D.; Grubelnik, A.; **Ruth, K.**; Panke, S.; Zinn, M. Quantitative analysis of bacterial medium-chain-length poly([R]-3-hydroxyalkanoates) by gas chromatography. *J. Chromatogr. A* 2007, 1143 (1-2), 199-206.

Ruth, K.; Grubelnik, A.; Hartmann, R.; Egli, T.; Zinn, M.; Ren, Q. Efficient production of (R)-3-hydroxycarboxylic acids by biotechnological conversion of polyhydroxyalkanoates and their purification. *Biomacromolecules* 2007, 8 (1), 279-286.

Ren, Q.; **Ruth, K.**; Thöny-Meyer, L.; Zinn, M. Process engineering for production of chiral hydroxycarboxylic acids from bacterial polyhydroxyalkanoates. *Macromol. Rapid Comm.* 2007, 28 (22), 2131-2136.

Ruth, K.; de Roo, G.; Egli, T.; Ren, Q. Identification of two Acyl-CoA Synthetases from *Pseudomonas putida* GPo1: One is located at the surface of polyhydroxyalkanoate granules. *Biomacromolecules* 2008. 9(6), 1652-1659.

Die VDM Verlagsservicegesellschaft sucht für wissenschaftliche Verlage abgeschlossene und herausragende

Dissertationen, Habilitationen, Diplomarbeiten, Master Theses, Magisterarbeiten usw.

für die kostenlose Publikation als Fachbuch.

Sie verfügen über eine Arbeit, die hohen inhaltlichen und formalen Ansprüchen genügt, und haben Interesse an einer honorarvergüteten Publikation?

Dann senden Sie bitte erste Informationen über sich und Ihre Arbeit per Email an *info@vdm-vsg.de*.

Sie erhalten kurzfristig unser Feedback!

VDM Verlagsservicegesellschaft mbH
Dudweiler Landstr. 99   Telefon +49 681 3720 174
D - 66123 Saarbrücken   Fax   +49 681 3720 1749
www.vdm-vsg.de

Die VDM Verlagsservicegesellschaft mbH vertritt

Printed by Books on Demand GmbH, Norderstedt / Germany